WHO'S AFRAID OF CHARLES DARWIN?

WHO'S AFRAID OF CHARLES DARWIN?

Debating Feminism and Evolutionary Theory

Griet Vandermassen

ROWMAN & LITTLEFIELD PUBLISHERS, INC.
Lanham • Boulder • New York • Toronto • Oxford

ROWMAN & LITTLEFIELD PUBLISHERS, INC.

Published in the United States of America
by Rowman & Littlefield Publishers, Inc.
A wholly owned subsidiary of The Rowman & Littlefield Publishing Group, Inc.
4501 Forbes Boulevard, Suite 200, Lanham, MD 20706
www.rowmanlittlefield.com

P.O. Box 317, Oxford OX2 9RU, UK

British Library Cataloguing in Publication Information Available

Library of Congress Cataloging-in-Publication Data

Vandermassen, Griet, 1970–
Who's afraid of Charles Darwin? : debating feminism and evolutionary theory / Griet Vandermassen.
p. cm.
Includes bibliographical references and index.
ISBN 0-7425-4350-1 (cloth : alk. paper) — ISBN 0-7425-4351-X (pbk. : alk. paper)
1. Feminism. 2. Human evolution. 3. Human biology. 4. Sociobiology. 5. Feminist theory. I. Title.
HQ1155.V36 2005
305.42—dc22

2004018264

Printed in the United States of America

∞™ The paper used in this publication meets the minimum requirements of American National Standard for Information Sciences—Permanence of Paper for Printed Library Materials, ANSI/NISO Z39.48-1992.

Contents

Foreword

GRIET VANDERMASSEN IS AN OPTIMIST. In this book, she hopes to engage feminist scholars and advocates in her mission to synthesize their writings with the science of evolutionary biology. It is evident that Vandermassen fervently believes in the value of an evolutionary perspective in understanding conflicts between the sexes. I endorse her belief. However, I am not such an optimist. Other feminist scholars and scientists have tried to bring these two heterogeneous perspectives together, with disappointingly little success. However, *Who's Afraid of Charles Darwin?* is the most balanced and scholarly attempt I have seen so far.

It is a major challenge to synthesize the feminist scholarship of those in the humanities with the science of evolutionary biology, not just because they differ in their accounts of femaleness and maleness. There is the more basic problem that the humanities and the life sciences differ in their methods and criteria for evaluating explanatory accounts of social dynamics. Much of humanist scholarship is focused on the impact and interpretation of events and ideas, or what psychologists might call the phenomenology of experience. It is not controversial that we define who we are by these feelings and experiences, but why do human beings feel and think the way they do? To the degree that the answers are panhuman rather than culturally particular, and they surely are to some degree, this sort of question is within the mandate of evolutionary biology. However, accounting for subjective experience has not been a part of that enterprise.

As this book reveals, the diversity among feminist scholars in their interpretations and claims about the origin and maintenance of male-female conflict may be of the same order of magnitude as the diversity of theories among life scientists. And even among biologists, there are those who disapprove of taking an evolutionary perspective to the study of people. My inclination is to consider how these criticisms can improve our science, not only because they may rightly highlight prejudice or oversight among scientists striving for objectivity, but also because criticisms, including misguided ones, provide windows on human values and information processing. It is personally satisfying to achieve a coherent view of aspects of the human condition and I infer that Griet Vandermassen feels the same. Vandermassen advocates embracing an evolutionary conceptual framework for the understanding of sexual conflict and human affairs more generally.

Some people have been wary about endorsing an evolutionary scientific perspective on the human animal because of a fear that evolved behavioral traits cannot be changed. That this fear is unfounded is revealed by everyday experience as well as by developments in medicine and the other sciences. For example, the structure of our teeth, our digestive physiology, and our appetites all indicate that *Homo sapiens* evolved as an omnivore, and yet people choose to be vegetarians with apparent ease. Of course, what someone eats is of relatively little consequence to others in comparison to such core feminist concerns as the subjugation of women. The worry here is that to suggest that the male psyche evolved to control women by direct action or by creating patriarchal institutions is to suggest that nothing can change. I protest. An evolutionary perspective on male psychology can result in a better understanding of the developmental conditions and environmental contingencies that regulate masculine psychological processes, and if there is a consensus that change is desirable, then such an understanding can result in better-informed approaches to child rearing and to the sociopolitical imposition of appropriate incentive structures.

My own research on men's use of violence against women has been guided by asking why men's motives and emotions should be such as to inflict harm on women they have loved and wanted. An evolutionary perspective on competition among males suggests an obvious hypothesis: that the human male psyche has evolved to treat sexual or romantic partners, to some degree, as alienable property. When positive inducements seem to be failing, coercive control is an effective tactic in many contexts, and it is little wonder that men who see women as scarce and contested resources often wish to limit their partners' autonomy. Considering both sexually proprietary inclinations and inclinations to use coercive tactics, a number of hypotheses can be made about the risk of men's violence against wives and romantic partners. Epidemiological analyses of violence against wives has revealed, for example, that young wives are at greater risk than older wives of being killed by husbands, wives in de facto marital unions are at greater risk than those in registered unions, and recently estranged wives are at greater risk than co-resident wives. These demographic risk markers are understandable from an evolutionary perspective, but they do not reveal the causal dynamics leading up to the fatal events, nor can they be used as predictors except in the crudest actuarial sense. Moreover, these results should not be interpreted as support for a male-centered patriarchal view of marital relations and institutions, because the violence itself is facilitated by such institutions. A masculine proprietary inclination to use violence against wives is amenable to change—we know this from cross-cultural variation in wife beating and in wife killing—but analysis of the social, economic, demographic, and historical correlates of such violence is one empirical approach to identifying its causes. Moreover, even in the absence of changes in men's inclinations to use violence against women and in the absence of change in institutional supports for a proprietary view of marital relations, empirical facts about risk markers can be used to increase the safety of women by focusing protection in high-risk circumstances.

Feminist advocates for the protection of women from violence and for women's autonomy have achieved goals that improve women's circumstances by increasing public awareness, by creating social and economic opportunities, and by imposing costs on perpetrators and supporters of the subjugation of women. However, they have not made significant contributions to a scientific understanding of male-female conflict. Indeed, such an understanding has been discouraged by the censoring of efforts perceived as antithetical to the political agenda and the ascendancy of their viewpoint.

Perhaps Griet Vandermassen's optimism about a synthesis of feminist and evolutionary approaches will be justified by the current cohort of young scholars and scientists who have benefited from the efforts of those who helped create a more balanced view of male-female relations. For readers who are curious about the history and content of what has sometimes been a bitter relationship between feminists and Darwinists, this book is for you.

Margo Wilson

Acknowledgments

I AM GREATLY INDEBTED TO JOHAN BRAECKMAN, who has stimulated me to undertake this research. His trust and advice have been of invaluable help to me.

I would also like to thank my colleagues at the University of Ghent for their support: Charlotte De Backer, Farah Focquaert, Koen Margodt, Karolien Poels, An Ravelinghien, Tom Speelman, and Walter Verraes. Thanks to David Barash, Simon Baron-Cohen, Milton Diamond, Terri Fisher, David Geary, Paul Gross, Nicola Koyama, David Schmitt, Robert Wright, and Ken Zucker for responding to my questions, and to Helena Cronin and Iver Mysterud for notifying me about indispensable literature. I am also grateful to Dylan Evans, Maryanne Fisher, Janet Radcliffe Richards, William Spriggs from Evolution's Voyage, and Ullica Segerstråle for sharing their insights with me, and to Tim Birkhead, Bobbi Low, and Elizabeth Meier, merely for expressing abundant enthusiasm about my line of research. Many thanks to Jim Brody, Marysa Demoor, Sarah Hrdy, Craig Palmer, and Margo Wilson for reading (parts of) the manuscript. Any remaining errors are entirely mine.

Most of all, thanks to Tim, for his love, trust, and encouragement.

Introduction

> Feminist analytical categories *should* be unstable—consistent and coherent theories in an unstable and incoherent world are obstacles to both our understanding and our social practices.
>
> Harding 1986:287, emphasis in original

WHILE HAVING CONSIDERED MYSELF A FEMINIST since my coming of age, it is only since a few years that I would use the description 'Darwinian feminist'. Before that time it just seemed self-evident that as a feminist, one did not trust what scientists had to say about the sexes. Science was, after all, a white, male enterprise. Did its history not testify enough to its inherent sexism, with Aristotle trying to prove female inferiority and Francis Bacon talking about "enslaving Nature"?

The scientific method itself was an androcentric approach to the world, used to dominate both women and nature, that much was clear to me, as it was to my fellow feminist (female and male) friends. We gathered in discussion groups, female-only as well as mixed (there was a lot of discussion about the pros and cons of both kinds of groups). Together we explored the ways in which masculinity and femininity were socially constructed, because we felt sure that everyone is born bisexual. Society, we knew, guided newborns into heterosexual pathways, in order to keep control of things. We considered strict heterosexuality and monogamy bourgeois, although many of us experienced how difficult it could be to break free from those conventions. Yet that only testified to the deep ways in which we had all been indoctrinated by society.

We found ourselves confronted with a few problems, however. Some of us obviously *liked* to behave in sex-typical ways. Now we girls could understand about some males wanting to act macho—after all, they got something out of it: dominance. What were we to do, however, with our desire to look feminine and behave in a feminine way? Was it permitted to shave one's legs or did that mean that one had given in to the demands of patriarchy? How were we to react when one of us said that she liked to sexually arouse herself and her boyfriend by playing hard to get, that most stereotypical of female tricks? Were we to condemn feminine behavior as a weakness, as a sign that one did not have the emotional strength to resist indoctrination? I felt that way. I felt guilty when trying to look pretty, because that implied I was weak. I felt bad when I didn't, because then I just felt unattractive.

We read and talked a lot, but we did not consult any dissenting literature. Why should we? We knew that we were right and that all the others were just defending their vested interests or were not enlightened enough to see the truth.

That is now all more than ten years ago, and in the meantime I have learned that it does not suffice for things to be true to just *want* them to be true. Reading scientific literature made me realize that my antiscientific stance of the past was mainly based on prejudice and ignorance. Through reading evolutionary biological and evolutionary psychological works it dawned on me that I had been misinformed by feminist writings on the subject. Moreover, I felt that evolutionary psychology could contribute to a better understanding of all the topics our gender groups had been discussing so elaborately (and so fruitlessly) in the past: patriarchy, gender identity and gender roles, mating preferences, and sexuality, without in the least being a defense of the status quo.

Evolutionary psychology, after all, only tries to describe how the evolved architecture of our minds reacts in a highly context-sensitive way to the environment it finds itself in. By changing the environment one can change the behavioral outcome, as the huge differences between cultures show us. That does not mean that men and women will ever become the same; there will always be *statistical*, mean-group psychosexual differences between them, just as there will always be differences between people. There is nothing wrong with that, it would seem, as long as people get equal chances to realize their potentialities. Since evolution is about variation and since evolutionary psychology is about statistics, moreover, no single man or woman should feel obliged to behave in a sex-typical way. Nor should they feel guilty if they want to do so—that is, of course, as long as they are considerate of the needs of other people.

In this work I want to reflect on all these issues. I want to show that evolutionary psychology and feminism, instead of being enemies, are in fact allies. Whereas the former can provide scientific insights into the ultimate causes of the 'battle of the sexes', the latter can draw upon these conclusions for the construction of a realistic program of social reform and can at the same time act as a watchdog for male bias in evolutionary theories. Since science is always socially embedded, awareness of different forms of bias is constantly needed. Sadly enough women's history of oppression has prevented them from entering the academic world until a few decades ago. They found the social and the biological sciences to be sometimes thoroughly male biased. Female scientists have been systematically correcting these androcentric views ever since; yet many feminist theorists remain distrustful of the scientific enterprise.

Elusive Definitions

Before embarking on this project it seems necessary to provide the reader with some definition of feminism. The question "What is feminism?" has, however, a staggering number of possible answers to it. To some, feminism is about equal rights and opportunities for both sexes. To others, it is about affirming women's difference. Some reject this notion of a female subject as essentialist and look for the ways in which the fragile construct of personal identity is supposedly constituted through discourse. Still others reject the "male" notions of logic, rationality, and ambition and embrace an intuitive, "holistic" worldview. The list of possibilities goes on and on. Theoretical inconsistency reigns to the extent that as a feminist, one can interpret this incoherence of theories as a deficiency or as a merit, depending on one's theoretical framework.

Since feminist theory is not, and has never been, a united body of thought, any attempt to define it will have to be so general as to tell us practically nothing, or it will have to encompass, in a nutshell, the main strands of feminism and their central premises. As feminist theorist Joan Scott says, whenever one tries to define feminism one runs into the problem that it has "only paradoxes to offer."[1]

To a newcomer in the field, this labyrinth of feminisms will probably seem bewildering at first. It may continue to do so for some time, until she has acquired enough background information to consider herself able to make a selection of the theories that seem to fit in with her knowledge and experience of the world.

That, at least, is the ideal situation. In practice still other motives often help to determine what strand of feminism she will choose to adhere to. Each feminism works with implicit assumptions of human nature, assumptions that more often than not are partially guided by political motives. On the one hand this should not surprise us, since feminism is in essence a political movement.

On the other hand one would expect people who want to change the treatment and position of women to desire reliable information about human nature, in order to act upon it as efficiently as possible. The variety of reasons for this frequently politicized stance toward what counts as reliable knowledge is one of the topics of this book.

But before starting to explore the field of tension between feminism, science in general, and evolutionary theory in particular, some description of the feminist movement is in order. Feminism, one might say, is the multiplicity of political and philosophical programs designed to explain and end sexist oppression. The history of feminism is rich, dramatic, and marked by rapid developments.[2] Its eldest and most influential strand, liberal feminism, emerged in the eighteenth century as a product of the Enlightenment. As such the liberal feminist focus was and still is on women as free and rational agents, who are entitled to the same basic rights as men (e.g., de Beauvoir 1949; Friedan 1963; Stanton, Anthony, and Gage 1881/1882; Wollstonecraft 1792; Young 1999). Consequently, liberal feminism is also called equity feminism. Its defenders tend to assume that the differences between the sexes are small and are mostly the product of socialization.

The nineteenth century also witnessed the birth of a feminist strand taking issue with the prioritizing of rationality that was typical of Enlightenment liberal theory. These 'difference feminists' highlight the differences between male and female qualities, arguing that the strengths of women may be a source of pride. They strive for a society that is more guided by female concerns and values than the kind of society we have now (e.g., Blackwell 1875; Fuller 1845; Gamble 1894; Gilligan 1982).

This first wave of feminism, the end of which is usually dated at about 1920, was followed by a second wave in the 1970s. The French existentialist philosopher Simone de Beauvoir started it with her famous book *The Second Sex* (1949), but the real explosion of feminist viewpoints came two decades later. May 1968, the burgeoning rise of civil rights and student movements, and the New Left: Women were active participants in them. Liberal feminism had been rearing its head again since the publication of Betty Friedan's *The Feminine Mystique* (1963), but now it was joined by a multiplicity of budding new strands, each starting from different theoretical premises.

A controversial new perspective was offered by radical feminism, which may be considered a radicalization of difference feminism. Some of the young women involved in radical politics in the 1960s became disillusioned by the sexist attitudes still prevalent in these movements. The liberation that was being fought for, they realized, was not meant to apply to the female sex: Males did not want to renounce their privileges. The condemnation of masculinity as essentially dominating and destructive and the celebration of femininity as essentially nurturing and egalitarian became the hallmark of the ensuing strand of radical feminism. So did the slogan "The personal is political": Politics had to be redefined to include personal relationships, since women's oppression not only was a case of unjust laws and unequal political representation, but also was deeply rooted in personal life. Radical feminists point to men's control over women's sexuality as the basis of women's oppression. They reject many of the values held in high esteem by liberal feminists, such as rationality, self-determination, and equal competition, because they deem them essentially male (e.g., Brownmiller 1975, 1984; Dworkin 1997; French 1992; Millett 1970). In this they deviate from difference feminists, who propose differences between the male and the female mind without valuing one above the other.

Most radical and difference feminists remain in the dark about the underlying causes of these differences. They seem to implicitly assume that a gender-typical psychology results from socialization pressures and learning experiences. This idea of difference is, however, met with suspicion by many other feminists, because it is a proposition that has long been used to keep women subordinate (e.g., Bleier 1985; Butler 1990, 1997; Connell 1995; Fausto-Sterling 1992; Harding 1986; Hubbard 1988, 1990; Tang-Martinez 1997; ten Dam and Volman 1995). When feminists of whatever strand tackle the issue of inherent gender difference, however—whether to endorse it or to criticize it—they tend to embrace a false nature-nurture dichotomy—a problem that misinforms a lot of feminist discussions and that will be analyzed in this book.

Another important feminist paradigm that took shape during the second wave is socialist feminism. Socialist feminists use traditional Marxist analysis to uncover the nature of women's subordination. They argue that gender disadvantages are, like those of class, structural and built into the very fabric of society. It is the patriarchal organization of labor, particularly under capitalism, that institutionalizes sexual inequality. Any analysis of the causes of women's subordination has therefore to be related to its socioeconomic context (e.g., Hartmann 1979).

Neither liberal feminism, nor radical or difference feminism, nor socialist feminism, however, have been able to explain just *why* men have always been in a privileged position, and exactly how the relationship between the sexes is historically reproduced. To some, psychoanalysis provides the answer. Derived from Freudian and Lacanian theory, the psychoanalytic view posits that girls and boys develop contrasting gender identities because they deal in a different way with the 'Oedipus and castration complexes' that arise in the early stages of normal psychosexual development. Freud disparaged feminism, and feminists have, in turn, long condemned Freud for devaluing women as morally inferior and for attributing women's misery to their unconscious desires. His description of infant development, however, provided a base upon which an important branch of contemporary feminist theory is built. Although a central component of second-wave feminist theory has been its rejection of what was called Freudian biological determinism, together with the obvious male bias inherent in Freudian ideas about female sexuality, a number of feminists have tried to revise these ideas from a feminist perspective. In 1974 Juliet Mitchell was the first to do so in her book *Psychoanalysis and Feminism.* The French philosophers Luce Irigaray, Julia Kristeva, and Hélène Cixous and the American psychoanalyst Nancy Chodorow have provided further influential feminist interpretations of Freudian and Lacanian theory.

During the past few decades, however, psychoanalysis has been established as missing every scientific basis. It is not based on experiments or clinical studies and it is refuted by modern neurobiological research. Virtually everything about psychoanalysis that lends itself to falsification has been falsified.[3] Although Freud has introduced partially useful concepts such as the unconscious, our current understanding of the workings of the brain indicates that looking through a psychoanalytic lens—in whichever updated version—is like studying the body using the ancient theory of the four bodily humors.[4] It is a nice theory, but it just isn't correct.

In the early 1980s, black feminism (e.g., hooks 1981), as well as other particularizing feminist voices such as lesbian feminism, started protesting against what they perceived as the false universality of the predominant feminist accounts of the plight of women. It was argued that those merely represented the experiences of white, middle-class, Western women, and that there could not be a universal category such as "women." There existed too many differences among women for that. It was because of this kind of binary thinking that phenomena such as sexism and racism could emerge, it was thought. By falsely universalizing the experiences of one's own group, the lives and needs of other people were overlooked. Surely there were as many differences among women as there were between men and women. Feminists had to stop searching for universal causes of women's oppression, since these causes would differ from context to context.

A useful refinement of feminist theorizing, this focus on interpersonal difference risked running into a paradox: If nothing distinguishes women as a group from men as a group, then how was their oppression to be explained, and why should they have to unite?

This conundrum is even more pregnant in postmodernist feminism. As an exponent of the third wave of feminism that emerged in the late 1980s, postmodernist feminism has moved away from an activist mode to a mode of academic theory and debate. It focuses on the subversion of the existing meanings of womanhood, on fractured identities, and on the rejection of all claims of universality:

> Consciousness of exclusion through naming is acute. Identities seem contradictory, partial, and strategic. With the hard-won recognition of their social and historical constitution, gender, race, and class cannot provide the basis for belief in "essential" unity. There is nothing about being "female," itself a highly complex category constructed in contested sexual scientific discourses and other social practices. Gender, race, or class consciousness is an achievement forced on us by the terrible historical experience of the contradictory social realities of patriarchy, colonialism, and capitalism. And who counts as "us" in my own rhetoric? (Haraway 1991d:155)

With this last question, historian of science Donna Haraway recognizes the contradiction at the heart of postmodernism: If the use of unified categories is to be condemned as essentialist and totalizing, then how can one speak meaningfully about the world and, indeed, how is a universalizing theory like postmodernism to be justified? What is the basis for developing political reform programs if there is only difference and if identities can only be unstable sites of shifting meanings? Haraway warns for the risk of "lapsing into boundless difference and giving up on the confusing task of making partial, real connection" (1991d:161), but she does not explain how this connection should be made, nor how it is even possible. Since in the postmodernist view objective knowledge is impossible and knowledge and power are always intimately connected, postmodernists seem to forsake any ideals of justice or equality as universal human goals.

Nancy Fraser and Linda Nicholson tackle this problem in their 1988 article "Social Criticism without Philosophy: An Encounter between Feminism and Postmodernism." As true postmodernists, they criticize theories "which claim, for example, to identify causes and constitutive features of sexism that operate cross-culturally." Such theories, they argue, "falsely universalize features of the theorist's own era, society, culture, class, sexual orientation, and ethnic, or racial, group" (1988:27). They consider feminist theories such as Nancy Chodorow's analysis of mothering,[5] purportedly producing women whose deep sense of self is relational and men whose deep sense of self is not, essentialist and ill-conceived. For a theorist to use "categories (like sexuality, mothering, reproduction, and sex-affective production) to construct a universalistic social theory is to risk projecting the socially dominant conjunctions and dispersions of her own society onto others," according to them (1988:31).

Consequently Fraser and Nicholson consider the question how to combine a postmodernist incredulity toward metanarratives with the social-critical power of feminism. The answer, according to them, lies in theory that is explicitly historical, attuned to the cultural specificity of different societies and periods and to that of different groups within societies and periods. Its mode of attention would be comparativist rather than universalizing, and it would dispense with the idea of a subject of history. In general, it would be pragmatic.

This answer, however, does not seem to solve anything. The same problem remains: If we are all just the products of societal and discursive practices, then there can be no objective standards for measuring human needs and human well-being, or for condemning sexist practices. Without some kind of universal ethical standard, based on scientific knowledge of human nature, cultural relativism is rampant.

The fundamental problems accompanying a radical social constructivism also trouble a variant of postmodernist feminism: queer theory. With the American philosopher Judith Butler as its most outspoken exponent, queer theory regards not only identities as discursively produced, but also bodies. It criticizes what it considers the normative heterosexuality of our society and of most feminist theorizing. Butler (1990, 1997) describes the body as constituted by the structures of language and politics. According to her, one cannot position oneself outside of those normative structures. Her philosophy is a combination of psychoanalytic (Lacanian) theory and the theories of the French postmodernist philosophers Foucault and Derrida. She rejects the notion of a 'subject'; all identities are products of discursive, normative practices that produce bodies and sexual desires. The performance of gender, the ritualized repetition of gender conventions, creates the illusion that gender has a kind of essence, but it has not. According to Butler and other queer theorists, sexual desire is basically ambivalent, but the heterosexual matrix of society makes us renounce our homosexual desires. A man learns to define himself as male by rejecting his femaleness, something that makes him desire the woman he will never be. She is his rejected identification.

Heterogeneity: Problem or Merit?

The heterogeneity of feminism can, of course, not be neatly classified. As Valerie Bryson (1999) points out, feminists frequently draw their ideas from more than one theoretical starting point, and these ideas often reinforce each other. At times they come into conflict, so that two contradictory sets of beliefs are held simultaneously. About this multiplicity of feminisms, evolutionary biologist and feminist Patricia Gowaty writes that they "most often arise organically as women (and some men) identify sources of oppression in their lives and struggle to free themselves" (1997b:2–3). It should be no surprise then that there are so many feminisms, she thinks. She continues: "It is also no surprise that divisiveness among feminists is one of the challenges of modern feminism. Each of our political philosophies probably deserves respect, because they have grown out of our lived experiences" (1997b:3).

I wholeheartedly agree with Gowaty's analysis of feminist divisiveness as being one of the challenges of modern feminism, but I think her conclusion is rather problematic. No matter how sincerely a political philosophy testifies to the lived experiences of the theorist, this does not in itself constitute any proof of the analysis being right. Consider radical feminist Andrea Dworkin's crusade against pornography. In *Life and Death: Unapologetic Writings on the Continuing War against Women* (1997) she provides an incisive account of the way her life has been shaped by rape and battering. One cannot fail to be moved by her rawly emotional writing style and by the misery she has been through:

> No one knew about battery then, including me. It had no public name. There were no shelters or refuges. Police were indifferent. There was no feminist advocacy or literature or social science. No one knew about the continuing consequences, now called post-traumatic stress syndrome, which has a nice dignity to it. How many times, after all, can one say terror, fear, anguish, dread, flashbacks, shaking, uncontrollable trembling, nightmares, he's going to kill me? At the time, so far as I knew, I was the only person this had ever happened to; and the degradation had numbed me, disoriented me, changed me, lowered me, shamed me, broken me. (19)

Nor can one escape feelings of moral outrage at the plight of (white, middle-class) young women only three decades ago or fail to appreciate the subsequent social changes due to feminism:

> It was a time when girls were supposed to be virgins when we married. The middle-class ideal was that women were not supposed to work; such labor would reflect badly on our husbands. Anyone pregnant outside of marriage was an outcast: a delinquent or an exile; had a criminal abortion or birthed a child that would most likely be taken away from her for adoption, which meant forever then. In disgrace, she would be sent away to some home for pregnant girls, entirely stigmatized; her parents ashamed, shocked; she herself a kind of poison that had ruined the family's notion of its own goodness and respectability. (Dworkin 1997:30)

Dworkin's analysis—and one of the basic tenets of radical feminism—that pornography is the root of female subordination, that it is "the DNA of male dominance" (Dworkin 1997:99) is, however, hardly borne out by the facts (e.g., Ellis 1989). Her account and that of the many women she has spoken with of how they have been hurt by pornography is very probably true. Yet our empathy with them must not lead us to far-reaching conclusions about power, sadism, and dehumanization working deliberately in pornography to establish the sexual and social subordination of women to men, as radical feminists assume. We can respect the theorist without having to accept the theory.

One problem with Dworkin's analysis is its implicit theory of human nature. When she writes about pornography that it is "what men want us to be, think we are, make us into; how men use us; not because biologically they are men but because this is how their social power is organized" (1997:99), women and men are presented as passive pawns in the social game, with women conditioned to be willingless victims and men to be eager woman haters. A similarly environmentalist conception of human nature underlies the work of many feminist theorists (e.g., Brouns, Verloo, and Grünell 1995; Butler 1990, 1997; Connell 1995; Dines, Jensen, and Russo 1998; Nicholson 1994). Theories like these not only cannot explain how present situations ever came to be (how have passive pawns ever been able to create complex social organizations?), they are at odds with modern scientific knowledge about an evolved human nature as well, as this book shows.

Although they seldom state them explicitly, each feminism has an undergirding theory of human nature. According to Patricia Gowaty (1997b), a profitable way to look at all the variety (and potential controversy) among feminisms is to look at the relationships among those theories of human nature. She considers them all proximate levels of analysis, that is, levels describing factors that mediate the expression of behavior (they describe the 'how' of behavioral expression, not the 'why'). Just like an endocrinologist, a geneticist, a sociologist, and a developmental psychologist might come up with different answers to the question of the cause of a given behavior and be all simultaneously right, the same might apply to the different feminisms, she says: Each describes part of the story. In addition, we can ask about the forces of evolution, about natural selection favoring this behavior or that, explanations biologists label the ultimate causes of behavior.

The stressful behavior of someone with a headache, for example, can be explained on different levels. On a sociological level, there may be a highly industrialized society with demanding jobs. On a psychological level, the person may be rather perfectionist and hence easily stressed. On a genetic level, he or she may have a genetically based tendency toward perfectionism. It is also known that headaches correlate with certain changes in the blood vessels in the neck and in the membranes that cover the brain. So a vascular theory of headaches is another way of understanding this person's condition. Then there is the neurological basis of pain in general, so a neurological theory of headaches is also available (Plotkin 1997). Finally, at an ultimate level, we can ask what benefits in terms of survival getting a headache in a stressful situation confers.

This 'multiple level of causations' point of view is, as Gowaty argues, certainly something that can help organize and facilitate the discussion among feminists. If one is aware that there are multiple causes of women's oppression, the variety of feminist political philosophies "can be—in theory—discussed without defensiveness about whether one is 'right or wrong'. Perhaps all are correct or partially correct—and we would all be better off for knowing that" (Gowaty 1997b:5).

The concept of multiple coexisting theories of human nature that all hit the mark seems, however, ill-conceived. No matter how many levels of description of human behavior there are, logic demands that there is only one 'human nature'. No matter how valuable ecofeminism, materialist feminism, or existentialist feminism—to name some other strands—might be for highlighting causes of women's subordination, as theories of human nature they inevitably clash. In order to be compatible as proximate explanations of behavior, they must by definition be grounded in a unified theory of human nature that can at the same time account for the impressive variation among individuals and among cultures. According to a growing number of scholars, this theory is the Darwinian theory of evolution by natural and sexual selection. As I try to show, an evolutionary point of view can help feminists solve many theoretical dilemmas. I claim that the confusing and often self-contradictory variety of feminisms not only has to do with the complexity of human behavior and of human social organization, but also with the absence of a deeply historical, evolutionary perspective. Such a perspective could serve to integrate the different feminisms. Sociologists Joseph Lopreato and Timothy Crippen argue that in its absence, "the sociology of 'sex roles' will continue to stray very far from a productive grasp of subject matter" (1999:169). The extent to which this rather stern-sounding condemnation is justified will be a matter of investigation in this work. It seems inevitable, however, that an evolutionary approach will do away with some much-cherished assumptions of what evolutionary psychologists John Tooby and Leda Cosmides (1992) call the 'Standard Social Science Model', the assumption that the human mind is virtually completely molded by the social environment.

In taking such an approach, one must nevertheless beware of the patronizing stance that Laurette Liesen detects for instance in David Buss's article "Sexual Conflict: Evolutionary Insights into Feminism and the 'Battle of the Sexes'" (1996). "One gets the impression that Buss believes evolutionary psychology can help save feminism from itself, making it more empirical and enlightened through evolutionary psychology's insights into the nature of relationships between males and females," Liesen, a political scientist studying the relationship between feminism and evolutionary psychology, writes (1998:111). Feminists have played an important role in elevating the scientific level of evolutionary theories of human and animal behavior, as this book makes clear, and probably continue to do so.

Sex and Gender

Another terminological issue that I have to touch upon before taking off is that of sex and gender. Feminist accounts of sex-differentiated behavior typically begin by making this distinction, although there is no agreement on the precise meaning of both concepts. Gender can be used in two very different and somewhat contradictory ways. In second-wave feminism a distinction was made between 'biological' sex and 'socialized' gender, with the latter referring to the internalization of sex-typical patterns of behavior. Since gender would explain the social construction of femininity, it would allow women to break out of the confines of what had been deemed their biological fate, it was thought (Segal 1999).

Gender is still often used as a contrasting term to sex, to depict that which is 'socially constructed' as opposed to that which is 'biologically given'. However, as Linda Nicholson says in her article "Interpreting Gender" (1994), the term has increasingly become used to refer to any social construction having to do with the male-female distinction, including those constructions that separate female bodies from male bodies: "This latter usage emerged when many came to realize that society not only shapes personality and behavior, it also shapes the ways in which the body appears. But if the body is itself always seen through social interpretation, then sex is not something that is separate from gender, but, rather, that which is subsumable under it" (Nicholson 1994:79).

Being a postmodernist, Nicholson, like queer theorist Judith Butler, defends this second usage. The position she would like feminists to endorse is that biology cannot be used to ground claims about women or men transculturally. About the distinction between sex and gender she writes:

> [T]his distinction, sex, was a word with strong biological associations. Early second-wave feminists correctly saw this concept as conceptually underpinning sexism in general. Because of its implicit claim that differences between men and women are rooted in biology, the concept of sex suggested the immutability of such differences and the hopelessness of change. To undermine the power of this concept, feminists of the late 1960s drew on the idea of the social constitution of human character. (Nicholson 1994:80)

Her criticism is clear:

> Many of those who accept the idea that character is socially formed and thus reject the idea that it emanates from biology do not necessarily reject the idea that biology is the site of character formation. In other words, they still view the physiological self as the "given" upon which specific characteristics are "superim-

> posed"; it provides the location for establishing where specific social influences are to go. The feminist acceptance of such view meant that sex still retained an important role: it provided the site upon which gender was thought to be constructed. (1994:81)

Nicholson cannot accept this 'biological foundationalism', as she calls it:

> The human population differs within itself not only in social expectations regarding how we think, feel, and act but also in the ways in which the body is viewed and the relationship between such views and expectations concerning how we think, feel, and act. . . . In this alternative view the body . . . becomes a variable rather than a constant, no longer able to ground claims about the male/female distinction across large sweeps of human history, but still there as always a potentially important element in how the male/female distinction gets played out in any specific society. (1994:82)

These quotations reveal her underlying motivation for accepting what counts as truth and what does not: It is clearly political. This attitude, as we see, resurfaces again and again in feminist thought. Nicholson's "Interpreting Gender" does not contain one single reference to modern scientific insights on the nature of males and females. This should not surprise us, since postmodernists put the entire edifice of knowledge into question and link scientific knowledge primarily to the exercise of power. As we see, however, not all strands of feminism take an antiscientific stance, and the feminist criticisms of and contributions to particular scientific disciplines often have helped to improve the scientific level of those disciplines.

Another interesting observation to be made about these quotations is that Nicholson—and a lot of feminists with her—automatically associates biological differences between the sexes with the hopelessness of social change. This association is definitely wrong, as many biologists and philosophers have shown. Still, the misconception persists.

Subsuming sex under gender has the advantage that it leads the outdated nature-nurture dichotomy to disappear. The way Nicholson proceeds is, however, untenable by any scientific standard. As I argue, feminism needs the biological and evolutionary sciences to settle this question (but the latter also need feminism, for different reasons). As long as feminists refuse to take into account scientific data about the sexes, (some strands of) feminism will stray. As long as there are feminists who, like the Dutch gender theorist Margo Brouns, are convinced that the question whether men and women differ by nature is "a *philosophical* question" (Brouns 1995b:47, my translation, emphasis in original), feminism will continue to find itself faced with contradictions.

In this work I use both concepts, sex as well as gender, to denote psychosexual and behavioral differences between men and women, whatever their presumed origin.

The Missing Link

This short—and inevitably caricatured—description of some of the main feminist traditions primarily serves to indicate that feminism is characterized by an immense multiplicity of theoretical, often contradictory perspectives. Although most of them can probably offer valuable partial explanations of gender differences at the proximate, direct level of behavior, none succeeds in providing an overarching explanatory framework. Rather than being wrong, they are somewhat myopic: They use the structural tendencies they find in society to explain patriarchy, but they leave these tendencies themselves unexplained.

Socialist feminism links the oppression of women to the class society, but why do humans constitute hierarchies in the first place? Radical feminism finds the origin of patriarchy in men's tendency to control women's sexuality, but they do not explain where this tendency comes from. Nancy Chodorow's psychoanalytic theory stresses the reproduction of mothering as a causal factor of male-female differences, but she cannot explain how this pattern of female mothering came to emerge in all known human societies.[6] Existentialist feminism links patriarchy to the male fear of 'the Other', liberal feminism links it to gendered socialization, lesbian feminism links it to the 'dogma' of heterosexuality, and ecofeminism sees the root of the patriarchal evil in androcentrism—the practice of placing male human beings at the center of one's worldview. The same problem remains: Their explanation needs another explanation in turn, one that can explain the ubiquitousness of the described behavior.

I believe that, in looking beyond the boundaries of Western and human societies and taking a perspective of humans as organic beings that evolved over evolutionary time, the problems that trouble feminist theories can be solved. If one applies an evolutionary framework, as Darwinian feminists do, one can integrate proximate feminist theories in the ultimate framework of evolutionary theory. Each level can guide discoveries and explanations on the other, thereby eliminating what proves to be unfounded. In taking this approach, an age-old question would be solved that Geert ten Dam and Monique Volman describe as "one of the most pressing questions of women's studies": "How is it possible that, generation after generation, people adjust themselves to the governing patterns of sex relationships?" (1995:160, my translation).

Plan of the Book

In order to shed light on the historical conflict between feminism and evolutionary theorizing, the first chapter starts by contextualizing the distrust that many feminists have of the biological sciences. Its focus is on past and present instances of misogyny in science, on the social embeddedness of science, and on answers to scientific relativism.

Chapter 2 dwells on the merits of and problems with feminist views of science, which can broadly be divided into empiricism, standpoint theory, and postmodernism. It asks whether science should be politically progressive and finally contends that feminist science as such does not exist.

Chapter 3 looks into the roots of sex differences as described and explained from an evolutionary perspective, starting with Darwin's writings on the subject and ending at the dawn of sociobiology in the 1970s. Feminist criticisms of several aspects of the theory of evolution by natural and sexual selection are critically dealt with as well as the impact of female scientists on theory formation in the biological sciences since the 1960s.

Chapter 4 analyzes feminist biophobia, the prevailing conception in gender studies departments that the psychosexual differences between men and women are almost solely the result of socialization practices. What are its sociological roots? What is the scientific evidence? I highlight the main erroneous reasonings of biophobian theorists: the nature-nurture dichotomy, the myth of genetic determinism, and the naturalistic fallacy, the idea that what *is* also *ought* to be.

In the fifth chapter I describe the rise of sociobiology, evaluating the feminist reactions to it. I set out the subsequent scientific developments leading to evolutionary psychology and explain the central premises of this discipline. The question of whether evolutionary psychology is scientifically defensible is tackled.

The last chapter is a plea for the integration of a Darwinian perspective into feminism. I explain why that is necessary and what evolutionary psychology has to offer feminists. I thereby dwell on the evolutionary origins of patriarchy. Finally I try to explain why Darwinism does not have the reactionary political implications that feminists fear it has. For as cognitive psychologist Steven Pinker says: "We should expose whatever ends are harmful and whatever ideas are false, and not confuse the two" (1997:48).

Notes

1. As cited in Segal 1999:200.

2. For an excellent introduction to the multiple feminisms, see *Feminist Debates* (Bryson 1999) or *Feminist Theory* (Donovan 2000).

3. For detailed refutations of psychoanalytic theory, see Crews 1996, 1998; Grünbaum 2002; Israëls 1999; Torrey 1992; and even a dedicated Jungian such as Storr 1989. Neurobiologist Eric Kandel (1999), firmly believing in the power of psychoanalysis, argues for an integration of neurobiology and psychoanalysis in order to provide the latter with a scientific foundation. Ironically, though, every neurobiological study that he refers to either refutes central psychoanalytic premises—regarding the suppression of memories, gender identity, and sexual orientation—or is in perfect agreement with standard psychological science. Nowhere does Kandel succeed in proving the usefulness of the psychoanalytic paradigm.

4. Black bile, blood, phlegm, and yellow bile. According to Galen, who developed this theory, these four fluids need to be present in equal quantity to be a healthy person.

5. Chodorow 1978.

6. As can be found in Brouns 1995c, structuralist psychoanalysts consider it a mystery that there are two sexes, no more, no less. It is rather amusing that it does not seem to occur to them that biology might offer some help in solving the mystery—although it must be said that there exist competing biological theories about the evolution of sex.

1

Science and Its Problems

> Science it would seem is not sexless; she is a man, a father and infected too.
>
> Woolf 2000:267

As women's traditional positions have often been justified by claims about a deep difference of nature between the sexes, it is not surprising that many feminists distrust scientific claims about biological differences between men and women. As neurophysiologist Ruth Bleier puts it, "biological theories have provided the scientific justification for ideologies that support, explain, mystify and obfuscate patriarchal relationships of power, domination and control" (1985:19).

The combination of a sad history of misogyny and prejudice in the history of science and the naturalistic fallacy in much of feminist thought probably does a fair job in explaining what I do not hesitate to call feminist biophobia[1]: the hostility to consider evidence from the biological sciences when it comes to elucidating gender difference. The naturalistic fallacy, the (wrong) idea that what ought to be is defined by what is natural, forces many feminists to deny the possibility of objective knowledge. They seem to reason that any unwelcome finding about the world has to be acted upon in a way that leaves no alternatives: "If science, however, simply uncovers 'truths,' can we criticize its theories for its *unavoidable political and social implications*?" (Bleier 1985:19, emphasis added). Bleier happily continues:

> That is a question that does not require tortured analysis because science does not simply uncover truths. Rather, science, a "system of cognitive production,"

> constitutes a "plurality of socially constructed ways of comprehending natural and social phenomena" (Mendelsohn, Weingart, and Whitley, 1977). A cultural product and institution, like literature, film, and political science, it is a body of interpretations and language used by persons who together create and discover meaning in what they study. And those persons, born into, reared, and situated within a particular class, gender, race, ethnic, and national context, have—like all other human beings—a world view. That view and life history of experiences and relationships, a scientist's values, beliefs, and biases, help to determine what questions scientists find interesting to ask, what assumptions they make, what language they use to pose questions, what they see and fail to see, how they interpret their data, what they hope, want, need, and believe to be true. (1985:19–20)

The various strands of feminism differ in their appreciation of science, as the next chapter shows. Liberal feminism believes in the possibility of obtaining knowledge that is objective and value-free, despite the obvious social embeddedness of science Bleier refers to in the second half of the quotation. The other feminist traditions tend toward some version of the cognitive relativism that is apparent in the first half of the quotation.

Which brings me to the subject of this chapter: The extent to which the distrust of science exhibited by many feminists (and others) is warranted. What do misogynist scientific theories and practices from the past tell us about the scientific enterprise itself? To what extent is objective knowledge possible? Does the social embeddedness of science sustain Bleier's claim that it is a cultural product just like film and literature? How can we escape scientific relativism?

Misogyny in Science: Facts about the Past

As many feminist writers have shown, the history of science, and of biology in particular, is characterized by prejudices about and the conscious or unconscious neglect of women and female animals in general.[2] I concentrate here on the history of theorizing about the sexes, as well as on the history of ethology (the study of animal behavior). For practical reasons the development of Darwinian evolutionary theory in particular is relegated to chapter 3.

It is not hard to find examples of overt sexism in the history of science and philosophy. The classic Greeks, like Aristotle, Galen, and Hippocrates, are rewarding subjects for showing how values and beliefs can determine theories. Hippocrates (460?–377 BCE) thought the uterus freely traveled through a woman's body, thereby giving rise to a number of physical, mental, and moral defects (the word "hysteria" is derived from the Greek *hystera* or uterus). He also claimed that menstrual blood was poisonous, because according to him women lacked the male capacity for driving away unclean substances through sweating. According to Galen (130?–?200 BCE), menstrual blood was a residue of blood in the food, something the small and inferior female body was not able to congest (Angier 1999).

Aristotle (384–322 BCE) "counted" fewer teeth in the mouths of women than in those of men—adding this dentitional inferiority to all the others kinds of female inferiority (Rosser 1992). He excluded women from serious consideration as moral entities. His claim that women, as totally passive beings, contributed nothing to conception but the womb as incubator reigned for ages. It led leading microscopists of the seventeenth and eighteenth centuries to assert that they had seen exceedingly minute forms of men with arms, heads, and legs complete inside sperm under the microscope (Bleier 1985). As Bleier notes, their observations were constrained not by the limited resolving power of the microscopes of the time but rather by Aristotle's 2,000-year-old concept of women.

The dichotomy between the male as active and the female as passive deeply resonated in philosophy. The German philosopher of the Enlightenment Fichte, for instance, claimed that it was completely rational for a woman not to strive for sexual satisfaction. Since female dignity consists in subjecting herself to the male through marriage, she cannot reasonably want to be free, he thought. Rousseau, Hume, Kant, and many other philosophers of the Enlightenment, too, in proclaiming the freedom and rationality of man (sic), showed themselves to have a rather restricted conception of human beings. Women, they thought, by nature did not possess the rational and moral qualities that would enable them to be active in public life (van Muijlwijk 1998).

According to historian Thomas Laqueur in his book *Making Sex: Body and Gender from the Greeks to Freud* (1990), the belief in the existence of two separate sexes did not emerge in the West until 1800. Before that time a one-body model reigned: the idea that the male was the primary form, with the female being a less perfect version of it. According to that model, the paradigm of anatomical thought during 2,000 years, women were imperfect men. Only around 1800, with the rise of modern science, did the one-sex model disappear and did female genitals get their own scientific denomination (Angier 1999; van Muijlwijk 1998).

This idea of the greater perfection of the male sex had for the first time been scientifically explained by Aristotle. His biology was based on the central premise that heat is the fundamental principle in the perfection of animals. Heat enables matter to develop. On the Aristotelian model, woman is colder than man, and this defect in heat will result in her brain being smaller and less developed than man's. To this Galen added the claim that because of the female defect in heat, her genitals could not emerge and project on the outside. According to him, the only difference between the genitals of the two sexes is that in women the various parts of them have remain stuck within the body, whereas in men they are outside the body (Tuana 1988). Nancy Tuana summarizes Galen's view as follows: "Woman remains, so to speak, half-baked" (1988:154). The history of anatomy attests to the power of Galen's analysis of woman's genitals, she says: "For centuries, anatomical drawings of women's internal genitalia would bear an uncanny resemblance to man's external genitalia" (1988:154).

The long tradition of a one-body model leads some feminists, such as Anthea Callen, to conclude:

> [T]here is no such thing as a "natural" body: all bodies . . . are socially constructed; all are representations which embody a complex web of cultural ideas, including notions of race, class and gender difference. Our own bodies, and our experience of them are also culturally mediated. (Callen 1998:402)

In analyzing eighteenth-century anatomical reproductions, Callen further notes that it is significant that female skeletons were rarely represented as embodying the *memento mori* theme. Women, as she points out, were commonly represented as not privy to knowledge: "The female skeleton bore the material, the reproductive, rather than the productive, spiritually or intellectually meaningful symbolism" (1998:412).

One might wonder whether the long reign of this particular one-body model warrants the conclusion that all bodies are socially constructed. Might it not be that those theories of the past were plainly wrong and that, because of the self-scrutinizing tendency of science, we have now developed better theories? A case to the point might be made, with one caveat, however: Without women to do some of the scrutiny, gender bias will not easily be eliminated. Indeed, the developments of the nineteenth and early twentieth centuries do not inspire much trust in the self-correcting nature of the scientific enterprise when it comes to matters of sex and gender. We would have to wait until the 1970s, when many women entered the academic world, for a thorough correction to start happening.

Nineteenth-century biologists and physicians argued on physiological grounds against the education of young women, for it might cause damage to their reproductive systems. Their ovaries and uteruses were claimed to require much energy and rest in order to function properly. Studying would deprive these organs of the necessary energy, resulting in their shrivelling away. The fact that poor women bore many children in spite of their working hard was not seen as a refutement of this argument. On the contrary, it was interpreted as a sign that these women were more animal-like and less highly evolved than upper-class females. Women were also held to be less intelligent than men, partly because of their smaller brains and partly because of the overall size differences between the sexes. It was thought that, since women cannot consume as much food as men, less food would be converted into thought, and that therefore men would always think more than women (Fausto-Sterling 1992; Hubbard 1990).

Elephants and whales undermined the brain size argument, however: If size determined intelligence, then these animals ought to be much smarter than humans. Dividing brain size by body weight provided no solution, since in this way women came out ahead. The French naturalist Georges Cuvier decided that intelligence could best be estimated by the relative proportions of the cranial to the facial bones. With this measure, however, birds turned out to be more intelligent than we are (Fausto-Sterling 1992). Later theories tried to locate the site of the highest mental abilities. For a time this was thought to be the frontal lobes, and soon researchers claimed that the frontal lobes of males were larger. When the parietal lobes were thought to be the location of intelligence, those of females were found to be shortcoming (Hyde 1996).

During the nineteenth century, philosophers and physical scientists in many European countries and in the United States devoted themselves to "proving" not only the inherent inferiority of women, but also that of people of color and Jews, using physiognomy, bumps on the skull, specious intelligence tests, and other contrived evidence (French 1992; Torrey 1992).

The mid-nineteenth-century medical view of women held that all their physical and mental problems derived from some sexual malfunction. In line with this, Victorian doctors taught that excision of the clitoris could cure a plethora of 'feminine weaknesses', including melancholy, hysteria, insanity, and epilepsy. Clitoridectomy was also used to eliminate masturbation, kleptomania, nymphomania, and lesbianism—actual lesbianism, a suspected inclination, or simply aversion to men. The procedure was used in Europe and the United States during the second half of the nineteenth century, in spite of its failure to cure the disorders it was intended to relieve. The last clitoridectomy known to have been performed in England to correct emotional disorder was done in the 1940s on a five-year-old girl (French 1992; Sheehan 1997).

Many examples of gender bias in the twentieth century can be found in anthropology and primatology. Alison Wylie (1997) cites one famous instance of androcentrism in anthropology. In the 1930s, Claude Lévi-Strauss entered one morning's experience in his field journal as follows: "The entire village left the next day in about thirty canoes, leaving us alone in the abandoned houses with the women and children" (1936, as cited in Wylie 1997:45).

Another example is the way in which women have been portrayed in reconstructions of human evolution in reference to subsistence and reproductive behavior in the 1950s and 1960s. The focus lay on "man the hunter," with a small, passive role reserved for women as mothers and mates. Only with the social climate of the 1970s encouraging questions about women's roles in evolution was the male centrality in social life challenged, with women revealed to be important as gatherers (and sometimes as hunters, too) and to have pivotal roles in social life (Zihlman 1997).[3]

Yet this neglect of women in anthropology is not only explained by male bias. It is true that the first half of the twentieth century figured some (very influential) women anthropologists such as Margaret Mead and Ruth Benedict (see chapter 4). However, at least until 1970 non-Western societies were mainly witnessed and interpreted by male observers, who indeed visualized relations between the sexes from a male perspective, but who were also often excluded in the field from areas of female activity and power. This observation of course underlines the critical importance of women researchers in the social and life sciences.

Historian of science Donna Haraway has documented how "primatologists tell stories remarkably appropriate to their times, places, genders, races, classes—as well as to their animals" (1991c:84). She points out how animal sociology has been unusually important in the construction of oppressive theories of what she calls "the body political" (1991a:11). Apart from being a sexist mirror of our own social world, the science of animal group behavior has also supplied legitimating ideologies of the patriarchal division of authority.

Haraway (1991a) describes how monkeys and apes were long seen as natural objects unobscured by culture. As such they would show the organic base in relation to which culture emerged. Animal sociology developed between 1920 and 1940, at a time when anthropology and sociology claimed to be able to understand humans solely with the aid of the concept of culture (see chapter 4). Still, animal societies were seen to have in simpler form all the characteristics of human societies, so one might learn from them how to organize human society. Thus animals retained an ambiguous place in the doctrine of the split between nature and culture.

In analyzing primatology, Haraway describes in particular the work of Clarence Carpenter, who established the skill of naturalistic observation of wild primates in the 1930s. His careful field studies were, however, "fully reflective of social relations based on dominance in the human world of scientists" (Haraway 1991a:15). Carpenter assumed, for instance, that the social field was organized by the dominance hierarchy of the males of the social group. In an experiment with rhesus monkeys he removed the "alpha male" (the animal judged most dominant) from the group and observed a subsequent increase in intragroup conflict and a loss of territory of the group relative to other groups. His conclusion was that dominance hierarchies are crucial in retaining social order. Without them, social order supposedly is seen to break down into individualistic competition.

Carpenter did, however, not do the control experiment of removing other animals than the dominant males, because this did not make sense within the whole complex of theories and unexamined assumptions of primates being models of human beings. Based on these assumptions, females were found to have a dominance hierarchy as well, but one of lower slope. They were primarily described from the point of view of their sexual relations with dominant males.

In current theorizing, dominance structures are still seen and examined, but other important causal explanations of functional organization have been added, such as matrilocality, female bonding, long-term social cooperation rather than short-term aggression, and flexibility rather than strict structures. A good case can be made that primatology has been transformed by the perspectives and critiques of women (see chapter 3).

When one looks at the history of science and philosophy from a female perspective, it is difficult to escape the feeling that all forces were united against women in order to handicap their development and to neglect their interests. But did misogyny really rear its ugly head at every level of the scientific enterprise?

Fables about the Past

As we see, the purported sexism in science sometimes solely resides in the eyes of the feminist beholder, especially when the scientific enterprise as such is under consideration.

Some feminists to a greater or lesser extent consider the scientific method itself an androcentric approach to the world, used to dominate both women and nature.[4] They regard scientific institutions, practice, and knowledge to be especially effective patriarchal tools to control and harm women in general. They trace the historical roots to the seventeenth century, to the "shift from an organic, hermetic approach to science (in which men revered and saw themselves as part of the environment and nature, and women as identified with nature) to a mechanistic, objective approach" (Rosser 1992:62). This glorification of reason and its resulting distance is then supposed to have justified men's domination and exploitation of women and the environment.

In elaborating this myth about an ancient organic worldview suddenly to be destroyed by scientific rationality, an accusing finger is often pointed at René Descartes, Francis Bacon, and Isaac Newton. In her article "The Cartesian Masculinization of Thought" (1986), philosopher Susan Bordo explains the epistemological anxiety of the first two of Descartes's *Meditations* as reflecting "anxiety over separation—from the organic female universe of the

Middle Ages and the Renaissance." She declares Cartesian objectivism to be "a defensive response to that separation anxiety, an aggressive intellectual 'flight from the feminine' rather than (simply) the confident articulation of a positive new epistemological ideal" (1986:248–249).

In a similar psychoanalytic vein, Evelyn Fox Keller, herself a biophysicist, associates modern science with a masculine fear of merging:

> Our early maternal environment, coupled with the cultural definition of masculine (that which can never appear feminine) and of autonomy (that which can never be compromised by dependency) leads to the association of female with the pleasures and dangers of merging, and of male with the comfort and loneliness of separateness. The boy's internal anxiety about both self and gender is echoed by the more widespread cultural anxiety, thereby encouraging postures of autonomy and masculinity, which can, indeed may, be designed to defend against that anxiety and the longing which generates it. . . . Along with autonomy the very act of separating subject from object—objectivity itself—comes to be associated with masculinity. (1982:239)

She sees this male fear of merging exemplified in the writings of Francis Bacon:

> To see the emphasis on power and control so prevalent in the rhetoric of Western science requires no great leap of the imagination. Indeed, that perception has become a commonplace. Above all, it is invited by the rhetoric that conjoins the domination of nature with the insistent image of nature as female, nowhere more familiar than in the writings of Francis Bacon. For Bacon, knowledge and power are one, and the promise of science is expressed as "leading to you Nature with all her children to bind her to your service and make her your slave." (Fox Keller 1982:242)

Fox Keller concludes, citing psychoanalyst Bruno Bettelheim, that "[o]nly with phallic psychology did aggressive manipulation of nature become possible" (1982:242).

Perhaps most outspoken is philosopher Sandra Harding, who famously called Newton's *Principia Mathematica* (1687) a "rape manual" (Gross and Levitt 1994) and who argues that "the most fundamental categories of scientific thought are male biased" (Harding 1986:290). Josephine Donovan also links the Newtonian paradigm to the suppression of "all which did not operate according to reason, according to mathematical principles of mechanism" (2000:19), including women.

These feminists often tend to argue for a return to the supposedly organic, holistic worldview that once reigned (Bordo 1986; Donovan 2000; Fox Keller 1982; French 1992). They share this desire with ecofeminists such as Carolyn

Merchant (*The Death of Nature*, 1980), whose basic tenet is that environmental oppression is linked to female oppression—a linking that, according to liberal feminist Rene Denfeld (1996), highlights an important aspect of current feminist thought. Like many feminists critical of science, ecofeminists believe that human beings existed for thousands of years in a peaceful, balanced, harmonious ecosystem, until the scientific revolution destroyed the harmony.

This mythical past, however, seems to owe its imaginary existence to wishful thinking, and putting the blame for the appearance of dualist thinking on Descartes, Bacon, and Newton is based on an incorrect reading of history. Martin Lewis (1996) points out that it is surely correct to consider the seventeenth century as a crucial turning point in intellectual history, but that the other (eco)feminist contentions do not stand up to scrutiny. The scientific revolution, he explains, was by no means a direct outgrowth of Descartes's or Bacon's beliefs about the relationship between people and nature:

> Francis Bacon no doubt inspired many scientists and engineers in the early modern period, but he was not a significant figure in the scientific revolution. Nor was the role of Descartes critical. The latter occupies an important place in the history of thought, both for his mathematics and for his insistence that philosophy must begin from a standpoint of utter skepticism. Most of his scientific views, however, were soon discredited. In actuality, Descartes' insistent rationalism and unacommodating mechanism thwarted the development of genuine science, for they discouraged empirical inquiry. (Lewis 1996:214)

Of course, Newton, like Galileo before him, was a pivotal figure in the scientific revolution. Moreover, his physics was certainly characterized by mechanistic explanations. That does not mean, however, that the scientific revolution—much less the Enlightenment—was predicated on a thoroughgoing metaphysics of dualistic mechanism. "What was significant was rather the formalization of a view of nature as characterized by regularity and by causal relations, and, more importantly, the development of reasonably reliable methods for answering specific questions about natural structures and processes" (Lewis 1996:214).

Metaphysical conceptions about the nature of life and about the relationship of humans to the rest of the natural world were of little consequence to the scientific revolution. Moreover, people have never needed metaphysical justifications for the exertion of power and control, nor for killing and warfare. To blame the industrial revolution's inventors of machine production for Cartesian dualism is hardly reasonable, since these people were mainly "practical men," with little interest in abstract philosophy or science, whose environment was still profoundly influenced by traditional Christian understanding (Lewis 1996).

What about the "organic female universe of the Middle Ages and the Renaissance," as Susan Bordo calls it? She is probably not talking about matriarchy, since feminists generally agree that institutionalized female dominance never existed (Fisher 1999; French 1992; Hrdy 1999a, 1999b; Low 2000; van Muijlwijk 1998). She seems to refer to a past many feminists do believe in: a supposedly peaceful, egalitarian past in which people lived in harmony with their environment. This scenario is projected onto the Middle Ages or onto prehistoric Europe, in the latter case completed with a cult of goddess worship. Contemporary tribal societies and peoples from the past are typically held up as eco-exemplars, proof to the essential human condition in the absence of corrupting Western ideas and practices (Denfeld 1996; Lewis 1996).

Archaeologists, however, strongly dispute the likelihood of a monolithic goddess religion, especially over thousands of years. Moreover, the evidence from the past indicates warfare, human sacrifice, violence, and depletion of natural resources. Jared Diamond (1997) describes how the colonization of new parts of the world by prehistoric man has always been followed by a wave of extinction of large mammals. Other than African and Eurasian mammals, who had coevolved with man for hundreds of thousands or millions of years and had thus learned to become afraid of him (or her), those animals were not afraid of man. They were as easy a prey as, millennia later, the dodo would prove to be. Were it not for the lack of a prehistoric philosophy of sustainability, the American continent would still accommodate elephants, lions, cheetahs, and camels. Diamond describes murder as a main cause of death in tribal societies, and he notes that even in so-called egalitarian societies there are always differences in power and prestige, be it that they are informal. Whenever the ecological circumstances permit it, the population will grow and formalized hierarchies will arise.

Data from anthropology show that primitive warfare is characterized by rape and the abduction of women, as it has probably always been. The food sharing of foraging peoples is driven by cost-benefit analyses, not by some intuitive communist feeling. They share when it would be suicidal not to do so (Low 2000; Pinker 1997; van der Dennen 2002). Behavioral ecologist Bobbi Low notes about sustainability:

> In many traditional societies . . . the "fruits of the forest"—berries, mushrooms, and so on—are common-pool resources: any villager can gather them. If the village is small and close-knit, if villagers have lived there a long time, if exploitative cheaters are easily detected and punished, and if there are no external markets, the villagers' use is liable to be shared for subsistence, and to be sustainable over long periods. But if the village has many transients, or if it is hard to catch and punish cheaters, or if anyone can convert forest products into lots of hard cash and move away, "take the money and run" defection is likely. (Low 2000:161)

Thus the problem of the depletion of resources is not new. It appears that it is a very human problem. To present but a few other eye-opening data: The Inca city Machu Picchu was probably deserted by its people because its extremely limited carrying capacity was exceeded by population and environmental pressures. Easter Island, with its famous statues, was deserted after its inhabitants committed ecocide: They transformed the densely forested island into a barren landscape (Shermer 2001). Before the population decline of the fourteenth century, many European societies were approaching local ecological limits and were suffering severe consequences (Lewis 1996).

In short, all the evidence indicates that there has never been an 'organic universe' to be destroyed. Similarly, the scientific revolution was never the monolithic key error of the West many feminists consider it to be.

Facts and Fables about the Present

No matter how overtly misogynist—or racist—some theories of the past may sound, this does not in itself prove anything about the supposedly sexist—or racist—character of particular contemporary scientific theories. It does not suffice to point to the past to prove that present-day theories cannot be but prejudiced. We see that it is often very difficult to decide if the purported sexism really deserves that name, or whether it is a case of feminist watchfulness working overtime.

Two forms of androcentrism or gender bias are commonly detected in the social and life sciences. First, the neglect of women, together with the treatment of masculine attributes and activities as typical of humankind. Second, the treatment of gender differences as a given.

The latter criticism will not be addressed here, since the topic of sex differences is interwoven throughout this work. As to the neglect of women, it is clear that some critiques hit the mark.

A much-cited example is Lawrence Kohlberg's famous and influential 1958 study of moral development in children, in which he argued that people go through a progressive list of moral stages, with the highest stage being the application of abstract, universal principles. According to Kohlberg, girls and women often do not reach this stage, because they "remain stuck" at the level of letting their moral reasoning be guided by personal relationships and social conventions. In her 1982 book *In a Different Voice*, psychologist Carol Gilligan argued that Kohlberg's study was fundamentally flawed by its dependence on samples comprised entirely of little boys (he followed the development of forty-eight boys for a period of over twenty years). She considered it a typical case of masculine standardization and showed, on the basis of her own research, that women do not think in a less moral way, but that they think differently. Their mode of moral reasoning is contextual, she argued. It centers around care, responsibility, and the needs of others, rather than around formal rights and rules.[5]

I suppose everyone will agree that Kohlberg's reliance on boys only for making a statement about human moral development was thoroughly androcentric, and that this study justly serves as an example of gender bias in science. It is less clear, however, whether Gilligan's equally influential study was more scientifically rigid than Kohlberg's. The anecdotal and impressionistic character of her methodology has been criticized by many academic psychologists, feminist and nonfeminist alike. Her claims are found to be lacking in acceptable empirical support (Hoff Sommers 1994, 2000). Indeed, Gilligan offers very little information on the three studies her thesis is based upon. She does not present any statistical analysis. She picks some of the answers of a small number of the 198 people she interviewed and does not tell us anything about the rest of them.

This situation would not be that bad if the studies themselves had been published and could thus be peer reviewed. They have not. They have never been available for public or professional review, something that makes philosopher Christina Hoff Sommers wonder whether Gilligan had *any* significant empirical support for her model. If so, she had a duty to present it, Hoff Sommers contends. "Research in the human sciences is notoriously difficult—all the more reason to present what evidence one has so that others can evaluate it and attempt on their own to replicate it in their research" (2000:107). According to her, other, independent research has not confirmed Gilligan's thesis that there is a substantial difference in the moral psychology of the sexes. The word "substantial" is of importance here. The sex differences that Gilligan describes seem to be much less pervasive than she herself suggests, but studies do tend to support a weaker version of her view: Women more than men prefer to think in terms of care (Campbell 2002). The debate about Gilligan's work continues.

In psychology and physiology, there is a long tradition of not using females as experimental animals, because that would add unwanted variation as females cycle (Zuk 1997). Needless to say, this produces incomplete science. Many feminists also argue that the predominance of male scientists has led to a bias in the choice and definition of the problems with which they have concerned themselves. Such claims abound with regard to the health sciences. The androcentrism is, however, not always as unequivocally clear as feminists presume it to be. Sometimes it is absent.

One complaint goes that contraception has not been given the scientific attention its human importance warrants and that, furthermore, the attention it has been given has been focused primarily on contraceptive techniques to be used by women (Fox Keller 1982).

Whereas it is probably true that it took a long time for contraception to be taken seriously by science, it is not clear, however, whether this was caused by androcentrism. Moreover, there are valuable scientific reasons for concentrating on women in the search for contraceptives, because of their different hormonal functioning. Furthermore, the answer to the question whether this focus on women is androcentric or not seems to depend entirely on the interpretative framework of the observer. Suppose the situation had been different and all scientific effort in finding contraceptives would have been directed at men. In that case one can easily imagine feminists raising the claim of androcentrism, because once again, women's fertility would be controlled by men. Similarly then, one can make the case for a female-friendly approach in the situation as it stands, since in this scenario women have their fertility in their own hands.

I am not arguing here that the research on contraceptives really *was* female-friendly. I just want to show that the situation is not always as clear as some feminists consider it to be. Take the example of menstrual cramps. As Fox Keller (1982) writes, many feminists argue that menstrual cramps, a serious problem for many women, has never been taken seriously by the medical profession. They presume that, had the concerns of medical research been articulated by women, this imbalance would not have arisen. At the same time, however, we find feminist evolutionary biologist Marlene Zuk (1997) complaining of the current tendency in the medical establishment and the media to treat premenstrual syndrome (PMS) as a disorder that requires attention and treatment in many women. One might as well say that PMS is finally taken seriously.

Then again there is the dubious discussion of women and acquired immune deficiency syndrome (AIDS). As Zuk (1997) points out, women have constituted an important part of the victims from the beginning. Yet they, and particularly prostitutes, have often been seen as "vectors" or "reservoirs" of the virus, of concern because of the likelihood of them spreading the disease to male clients or other sexual partners, rather than as persons with AIDS in their own right. She notes that, as with other sexually transmitted diseases, AIDS is too often seen as something women give to men, with men the real victims.

In her book *Ceasefire!* (1999), liberal feminist Cathy Young weighs the evidence of male bias in medicine in the United States. She concludes that medical research indeed has blind spots where women are concerned, particularly in areas stereotyped as male. For instance, more than 30 percent of studies of alcoholism have been limited to men, while fewer than 6 percent have focused only on women. She also mentions AIDS as another area where female-specific problems did not get enough attention for a long time—an imbalance that, according to her, has now been corrected: "By the mid-1990s, though, women were *overrepresented* in an NIH-funded[6] AIDS clinical trial: They made up over 30 percent of the subjects, even though they accounted for about 12 percent of AIDS cases" (Young 1999:77–78, emphasis in original).

Despite some obvious imbalances in past or current research, however, Young warns for the mistake of blaming every discrepancy in research on sexism. A 1988 American study of the role of aspirin in heart attack prevention, for instance, has become a symbol of male bias in medicine: 22,000 men and not a single woman were tested. As a (woman) epidemiologist explained to Young, however, in terms of getting an answer relatively fast, it made sense to do the trial this way, since heart disease is much more frequent in men than in women. In order to keep the sample at a feasible level, it made sense to look at a group in which one can expect a fairly high rate of heart attacks, and then to decide if a similar study needed to be done in women.[7] Young also shows how some of the discrepancy reflects legitimate priorities: Since men die of heart attacks three times as often as women before the age of sixty-five, greater attention to them is justified. The same logic applies to other diseases that affect the sexes to different degrees.

Some feminist claims are, regrettably, false or deeply misleading. Take Marilyn French's accusation in *The War against Women*, that "[w]oman-specific diseases like breast and ovarian cancer have not been studied as thoroughly as male-specific diseases like prostate cancer and are more likely to be fatal. . . . *Only 13 percent of the dollar 7.7 million NIH budget is spent on women's issues*" (French 1992:134, emphasis in original).

Yet as Young (1999) shows, from 1981 on the United States has spent much more money on breast cancer research than on prostate cancer research—a discrepancy that again reflects legitimate priorities, since prostate cancer kills on average ten to fifteen years later than breast cancer. As to the NIH budget, what French does not mention is that "NIH funding for male-specific health projects accounted for just *under 7 percent* of its total expenditures; 80 percent of the money went to study diseases that afflict both sexes" (Young 1999:76, emphasis in original).

The list of examples of real or supposed sexism in science is long. It makes clear that the scientific enterprise is socially embedded and that therefore we should always be aware of possible forms of bias, in traditional scientific research as well as in feminist research.

The Social Embeddedness of Science

The history of science offers countless examples of the way in which scientific perception of reality is influenced by expectation. Using the words of William

Whewell, a nineteenth-century philosopher of science, Marcy Lawton, William Garstka, and J. Craig Hanks have called this "the mask of theory," a mask that "has always been shaped by culture and history" (1997:63):

> [T]he "givens" of any research program include assumptions, conceptual constructs, and reifications that may not be based in fact, but serve as heuristic devices that may limit the field of view offered by the ever-present mask of theory. Thus, the mask of theory, which none can put off, sometimes makes it hard for us to see the face of nature, even after we have had a good look. The full meaning of the data we collect may not be apparent to us, because the theoretical framework within which we ask questions constrains what we see and how we interpret it. (63–64)

They detect at least two ways in which theory may obscure reality. First, it may limit what we choose to look at. The study of cooperative breeding in birds offers one such example. As Lawton, Garstka, and Hanks describe, the phenomenon of birds caring for nestlings that were not their own offspring instead of breeding on their own was hardly acknowledged from the 1930s until the 1960s. Although there were some observations, they were seen as anomalous and explained away. With the emergence of kin theory in the 1960s and 1970s (see chapter 3), cooperative breeding became theoretically possible and helping at the nest was discovered in hundreds of bird species.

Second, the mask of theory may limit what is *seen*. The authors here offer three examples of the way in which gender bias is woven into the mask of theory, and how it has the power to distort our gaze largely because it is invisible. I elaborate only one of their examples here: that of gender bias in the scientific study of dominance relationships in the pinyon jay, a highly social member of the crow family. Lawton, Garstka, and Hanks (1997) concentrate on the volume *The Pinyon Jay* by Marzluff and Balda (1992). As they describe it, "even after watching pinyon jays *not fight* for 20 years, the authors were constrained by theoretical expectations to find what 'everyone knows' is the case; that is, social organizations are built around male dominance hierarchies" (1997:69, emphasis in original). To find these 'alpha males' required some experimental ingenuity, however, because adult male pinyon jays almost never fight. It was determined that a "dominant" bird had "aggressed" against a "submissive" bird when the former turned its head and looked at the latter. *Female* birds were, however, displaying fierce fighting in late winter and early spring. These behaviors were left out of the analysis of dominance relationships for this species, although strong clues that the behavior of females might influence the so-called male hierarchy were present. The reason was that the authors of *The Pinyon Jay* had not been open to explanations other than that which their theory demanded.

Another example of the mask of theory is offered by sociobiologist Sarah Hrdy in her important book *The Woman That Never Evolved* (1999a, first edition 1981). Infanticide in great apes by males unrelated to the infant was long regarded as a pathological aberration, since in the early days of ethology nonhuman primates were considered peaceful, group-oriented species. By closely observing one particular species, the Hanuman langur, Hrdy was able to document that infanticide is a recurring event and that it actually serves the reproductive interests of the male. By eliminating a baby unlikely to be his own, the male reduces the number of offspring sired by his competitor. Because the mother lost her baby she becomes sexually receptive again, and the murderous male can then inseminate her himself.

It took her discovery and her explanation of infanticide in primates some time to become accepted by primatologists and social scientists. She notes: "The history of our knowledge about primate infanticide is in many ways a parable for the biases and fallibility that plague observational sciences: we discount the unimaginable and fail to see what we do not expect. . . . I believe that one of the important factors determining whether infanticide was recorded or missed may have simply been expectation" (Hrdy 1999a:89).

The Woman That Never Evolved was such a ground-breaking work mainly because it was the first reconciliation of feminism and evolutionary theory to be published. With it Hrdy wanted to "convince fellow sociobiologists that we had to widen our field of vision to include the interests and perspectives of both sexes if we were to have a comprehensive understanding of the evolutionary process" (1999a:xvi). At the time, females were still stereotyped as passive and coy. Through the expansion of the theoretical framework by female authors such as Hrdy, females were often found to behave in unexpected promiscuous ways (see chapter 5).

Women scientists have transformed the natural sciences, as Hrdy (1999b) writes. Whereas anthropology, primatology, and evolutionary theory were very male centered until the 1970s, the changing genderscape in the academy since then, with its broader inclusion of women, brought new emphases and new topics of study, such as female-female competition and the active role of mothers in evolutionary processes. Before that time, primatologists and those social scientists that relied upon their work focused mainly on the supposedly greater competitive potential of males, the importance of males in structuring the society, and the apparent inability of females to maintain stable social systems. Their models turn out to have been projections of patriarchal fantasies.

Female perspectives have now become part of evolutionary science, and of science in general. In many cases it took female or feminist scientists to make visible data that had always been there, but that had never been recognized by male scientists (see chapter 3). Androcentric theories and hypotheses probably still abound, but so do gynocentric theories and hypotheses. The "right" to gender bias has now been extended to both sexes.

The point is that, since we are a two-sexed species, individual bias is probably inevitable in the study of animal and human behavior. As evolutionary psychologist Geoffrey Miller puts it in his epilogue to *The Mating Mind*:

> One's understanding of human sexuality and human behavior depends, to some extent, on one's sex. Throughout this book, I have tried to write first as a scientist, second as a human, and only third as a male. Yet some of my ideas have probably been too influenced by my sex, my experiences, and my intuitions. The trouble is, I don't know which ideas are the biased ones. . . . Perhaps others will be kind enough to identify them. A woman might have written a book about mental evolution through sexual selection with different emphases and insights. Indeed, I hope that women will write such books, so we can triangulate on the truth about human evolution from our distinctive viewpoints. (Miller 2000b:427–428)

Part of the solution to the problem of bias in science thus seems to lie in the collective nature of the scientific enterprise: Men and women can correct each other's biases, just as, supposedly, age, ethnicity, and class will also provide different perspectives. But does this not at the same time amount to a case for scientific relativism? How do we know that we have incorporated "enough" perspectives for bias to be eliminated? And does not the history of science prove that scientific knowledge is "just a social construct," changing through the ages, depending on the cultural climate?

Answers to Scientific Relativism

Part of what was once considered scientific knowledge does not meet our present-day standards of scientificity and has hence been rejected. I want to show that this observation does not warrant the scientific relativism or social constructionist epistemology that abounds today in feminism, in philosophy, and in the sociology of science. To cite but a few examples: "We do not have facts in psychology—rather, we have theories and scientific data, but we sometimes start thinking that we have facts. . . . Constructionism . . . argues that people—including scientists—do not discover reality; instead, they construct or invent it based in part on prior experiences and predispositions" (Hyde 1996:107). "Perhaps 'reality' can have 'a' structure only from the falsely universalizing perspective of the master" (Jane Flax, as cited in Harding 1986:294). "The natural world has a small or non-existent role in the construction of scientific knowledge" (Harry Collins, as cited in Cole 1996:275).

It is clear that these authors think that truth is but an illusion, or worse, that it is a power grab. Is that so? Consider Aristotle's claim about women being imperfect men, which has definitively turned out to be both an illusion *and* a power grab. Does this mean that social constructionism is right and that to call a statement true "is just to give it a rhetorical pat on the back," as postmodernist philosopher Richard Rorty (as cited in Haack 1996a:57) puts it?

Philosopher Susan Haack (1996a) points to the ubiquitous confusion between truth and truth *claims.* One can say that there are many truths in the sense of there being different but compatible descriptions of the world that are all true. But if one thinks that different and incompatible descriptions of the world may be all true, one is talking about truth *claims.* What passes for known truth is often no such thing, Haack contends. That does not mean that the idea of objective evidence is just ideological humbug:

> "True" is a word that *we apply* to statements about which we agree, simply because, if we agree that p, we argue that p is true. But we may agree that p when p is *not* true. So "true" is not a word that *truely applies* to all or only statements about which we agree; and neither, of course, does calling a statement 'true' mean that it is a statement we agree about. (Haack 1996a:61, emphasis in original)

According to Haack, the genuine inquirer is motivated

> to seek out and assess the worth of evidence and arguments thoroughly and impartially; to acknowledge, to himself as well as others, where his evidence and arguments seem shakiest and his articulation of problem or solution vaguest; to go with the evidence even to unpopular conclusions or conclusions that undermine his formerly deeply held convictions; and to welcome someone else's having found the truth he was seeking. (1996a:58–59)

Some will probably argue that this ideal of "assessing the worth of evidence impartially" is unattainable in practice, and they will no doubt be right. But one can at least *try* not to reason in a way that is already made evidence- and argument-proof by one's commitment to a proposition identified in advance of inquiry and hope that sooner or later one's own biases will be corrected by one's peers.

Haack (1996b) thoroughly debunks the claim that science is *simply* a social construction. From the true observation that scientific inquiry is a social enterprise one can come to the ambiguous conclusion that scientific knowledge is 'socially constructed'. But to equate this with the conclusion that scientific knowledge is nothing more than the product of processes of social negotiation is false and misleading. As Haack argues, the processes by which scientific knowledge is achieved are processes of seeking out, checking, and assessing the weight of evidence.

> What is distinctive about inquiry in the sciences is, rather: systematic commitment to criticism and testing, and to isolating one variable at a time; experimental contrivance of every kind; instruments of observation from the microscope to the questionnaire, sophisticated techniques of mathematical and statistical modeling; *and* the engagement, cooperative and competitive, of many persons, within and across generations, in the enterprise of scientific inquiry. (Haack 1996b:261, emphasis in original)

It is a misunderstanding, she points out, that the warrant status of a scientific claim is "just a matter of social practice." How justified a scientific community is in accepting a specific claim does not depend on how justified they *think* they are, but on how good their evidence is. It is also a misunderstanding to think that

> "how good," here, can only mean "how good relative to the standards of community C," since standards of evidence, it is supposed, vary incommensurably among communities or paradigms. But the supposed community—or paradigm—relativity of evidential standards is not real incommensurability, but deep-seated disagreement in background beliefs, producing disagreement about what counts as relevant evidence. (Haack 1996b:262–263)

As an example of background beliefs she gives the belief in graphology. Suppose, she says, A thinks this candidate for the job should be ruled out on the grounds that his handwriting indicates that he is not to be trusted. B thinks graphology is bunk and scoffs at A's "evidence." Their judgments of relevance of evidence depend on their background beliefs, but relevance of evidence is, nevertheless, objective.

As we see in the next chapter, this disagreement in background beliefs is also of crucial importance in feminist conceptions of science.

Finally Haack (1996b) analyzes an analogous kind of confusion: the claim that the objects of scientific knowledge are socially constructed. Scientific theories are of course devised and articulated by scientists, she argues. Theoretical concepts like electron, gene, force, and so forth could be considered their creation. But it does not follow, and neither is it true, that electrons, genes, forces, and the like are brought into existence by the intellectual activity of the scientists who create the theories that posit them. It is true that, as science proceeds instrumentation and theory get more and more intertwined, and one increasingly encounters claims that refer to what one might call 'laboratory' phenomena. But that such phenomena are created in the laboratory does not mean that they are made real by scientists' theorizing. Nor does the fact that we cannot describe the world without using language prove that reality is socially constructed; one does not follow from the other.

In criticizing social constructionism, physicist Loren Fishman (1996) notes that the arguments of the constructionists depend, essentially, on the view that feelings—in particular, feelings generated by social demands—can generate beliefs of the kind that scientists typically adhere to. In the constructionist view, one believes because social pressures force one to believe. A belief is, however, not just a proposition that is assented to in some haphazard fashion, as Fishman points out. Rather, it is a deep, intricate, and finely coordinated mental structure that bears the traces of complex inference. These characteristics are an important part of what makes scientific beliefs scientific. One cannot believe just because one wants to, any more than one can forget or be happy by wilful act.

Of course, social factors limit the evidence available to an inquirer and direct his or her attention to one class of questions rather than another. This kind of "construction" is more or less universally conceded. "But that beliefs, in and of themselves, with their necessary mechanisms of justification standing at the ready, are adopted in toto independently of internal logical coherence and with indifference to observable evidence, out of the need or desire to conform with some kind of shadowy social ethos, defies plausibility" (Fishman 1996:92).

There are also potential hindrances to scientific knowledge due to the organization and environment of science: pressure to find evidence supporting a politically desired conclusion or to ignore questions perceived as socially disruptive; pressure to solve problems perceived as socially urgent; the necessity to spend large amounts of time and energy on obtaining resources; dependence for resources on bodies with an interest in the research coming out a certain way, and so on (Haack 1996b; Stengers 1997).

Because of this dependence of the scientific enterprise on social, political, and industrial interests, science is seldom "pure" in the sense of "completely disinterested." It does not follow, however, that knowledge cannot be objective. Disinterestedness and objectivity are not synonyms. It is possible for a scientist to do objective research on a topic for deeply personal, subjective reasons. And even if the passions and ambitions of individual scientists may stand in the way of objectivity, and despite the fact that bias sometimes reigns, the communal enterprise of science as a whole, with its inherent self-criticism, its high standards of proof, and its ability to generate new insights in the end, leads to knowledge that is relatively trustworthy.

Sociologist of science Stephen Cole (1996) writes that social constructionists have never been able to show that social processes have influenced the actual cognitive content of some piece of science rather than the foci of attention or the rate of advance. While this may ultimately be true, it is also somewhat misleading. The supposition that female primates are not competitive nor sexually assertive, for instance, has long been part of the "actual cognitive content" of evolutionary science, until women primatologists started to correct this bias in the 1970s. This suggests not only that science is subject to bias and self-correcting, but also that it can take very long for this self-correction to take place if bias is deeply ingrained—something that the history of theorizing on the sexes makes clear.

But despite its flaws at times, science is still the best method ever devised for understanding reality, the best knowledge filter ever invented. The scientific methods developed over the past four centuries were specifically designed to help us avoid errors in our thinking. The fact that scientists, just like everyone else, are prone to find what they are looking for is the very reason that science is constructed to be self-correcting. It has built into it methods to detect both conscious and unconscious bias, and this is what sets it apart from all other knowledge systems and intellectual disciplines. "If it were not for this self-correcting mechanism, in fact, science could not have made the remarkable progress it has over its 500 year story" (Shermer 2001:317). Science's greatest weakness is its greatest strength. Others concur: "Improvement is not guaranteed. But science shows it more reliably than most human activities" (Cronin 1991:4). "Scientific method can be the antidote to hidden agendas" (Campbell 2002:22).

Loren Fishman's conclusion is analogously optimistic:

> The influence of cultural factors on thought and feeling is undeniable. The power of culture to dictate which facts we focus on, which conclusions we are most interested in, is all too apparent. Yet all facts are equally accessible to logic. The consistency of the principles of science is what holds science together. The characteristics of a society may hold its science together—for example, through practical respect for the scientific method, widely applied standards of measurement, a cultural commitment to free speech and telling the truth, and so forth. But in general the idiosyncrasies of factious, polymorphous society are not part of science. (1996:94–95)

Liberal feminists will agree with Shermer, Cronin, Fishman, and Campbell, whereas many feminists from other strands will tend to raise objections of various kinds. The various feminist conceptions of science are highlighted in the next chapter.

Notes

1. A term proposed by evolutionary psychologists Martin Daly and Margo Wilson (1988).

2. For example, Angier 1999; Birke 1999; Bleier 1985; Brouns 1995a; Blackwell 1875; Cronin 1991; Fausto-Sterling 1992; Fedigan 1997; Fox Keller 1982; French 1992;

Haraway 1991; Hrdy 1997, 1999a, 1999b; Hubbard 1990; Hyde 1996; Rosser 1992; Schiebinger 1999; van Muijlwijk 1998; Zihlman 1997; Zuk 1997, 2002.

3. Anthropologist Adrienne Zihlman maintains in this article that in the 1990s "the 'Second Sex' continues to be portrayed as the handmaidens of society's male players," despite the "increasingly powerful current of research demonstrating the centrality of women and female primates in social life and the evolutionary process" (1997:109).

4. For example, Bleier 1985; Bordo 1986; Fox Keller 1982; French 1992; Haraway 1991e; Harding 1986; Hubbard 1990; Rosser 1992; van Muijlwijk 1998.

5. Bobbi Low notes that adults who live in small, traditional, kin-based groups with competition from outsiders, also scored relatively "early" on Kohlberg's scale. She comments: "I am far from certain that I would assume from this a certain moral 'progression' as Kohlberg did. The ecology of social behavior is far different when you live among your relatives and the same friends all your life, than when you spend your life in a large and fluid society, and the rules should also be different" (2000:167).

6. National Institutes of Health (NIH).

7. Cathy Young does not tell us whether a similar study was indeed proved to be needed in women. She does mention that in 1988, a less rigorous study with nurses recording their own aspirin intake had been underway for three years. Londa Schiebinger (1999) refers to this study. It involved following 87,000 nurses for six years to study the correlation between taking aspirin and the risk of heart attack. However, Schiebinger does not mention the results of the study.

2

Feminist Views of Science

> I am suggesting that a feminist science practice admits political considerations as relevant constraints on reasoning. . . . [I]f faced with a conflict between [political] commitments and a particular model of brain-behavior, we allow the political commitments to guide the choice.
>
> Helen Longino, *Science as Social Knowledge*, 1990
> (as cited in Koertge 1996a:271)

As it is the purpose of my work to analyze the often conflicting relationship between feminist and evolutionary theory, it will be necessary to investigate the various feminist views of the scientific enterprise more in detail. Comprehension of the underlying conceptions of science is needed to understand fully the vigor of many feminist attacks on the application of evolutionary theory on human behavior and the human mind, as happens in sociobiology and evolutionary psychology. I thereby make use of Sandra Harding's (1986) categorization of feminist epistemological positions into (traditional) empiricism, (fairly radical) standpoint theory, and (very radical) postmodernism. As the focus of this book is on feminism, it does not discuss other (nonfeminist) critiques of science.

In my mind the continuum from traditional to radical positions at the same time represents a continuum from constructive science criticism, with viable alternative hypotheses and experiments being offered, to a destructive attack on the scientific enterprise as a whole. Since the boundaries are vague, my attribution of particular epistemologies to particular feminist traditions and theorists can only be tentative.

It shows that radical science criticism is riddled with contradictions and that most feminist contributions to science are ultimately grounded in what Sandra Harding calls an empiricist position, that is, traditional scientific methodology. This observation leads me to a further elaboration of a question already touched upon in the former chapter: the question whether there are really so many different ways to do science. Is the incorporation of female perspectives in explanatory models not merely inherent to the maturation of a scientific discipline? And hence, does something like feminist science really exist?

Empiricism: The Belief in Gender-Free Science

The empiricist position holds that good scientific research is not conducted differently by men and women. It recognizes that no individual can be neutral or value free, but it assumes that the scientific method, when properly used, permits unbiased research due to the collectivity of the scientific enterprise. It suggests that now that the bias of gender has been revealed by feminist critiques (such as the exclusion of females as experimental subjects and the focusing on problems of primary interest to males), scientists can take this into account and correct it. Empiricism is embraced primarily by liberal feminists. The liberal feminist combination of the belief in the validity of the scientific method and the typical assumption that there are no inherent gender differences also implies that once social barriers are removed and women are no longer discriminated against on the basis of sex, they will constitute 45 percent of scientists. That is, after all, their proportion in the overall workforce population. No changes in science itself are needed (Rosser 1997).

Sociobiologist Sarah Hrdy considers herself a liberal feminist, albeit one whose views on the sexes are informed by a Darwinian perspective (personal communication, 2001). Hence she trusts that research on sex differences will be able to rid itself of its problems. As she puts it:

> There are of course antidotes to the all-too-human elements that plague our efforts to study the natural world. Common sense in methodology is one. No one will ever again be permitted to make pronouncements about primate breeding systems after having studied only one sex or after watching only the conspicuous animals. A recognition of the source of bias is another. If, for example, we suspect that identification with same-sex individuals goes on or that certain researchers identify with the dominant and others with the oppressed and so forth, we would do well to encourage multiple studies, restudies, and challenges to current theories by a broad array of observers. We would also do well to distinguish explicitly between what we know and what we know is only interpretation. But

> really (being generous) this is science as currently practiced: inefficient, biased, frustrating, replete with false starts and red herrings, but nevertheless responsive to criticism and self-correcting, and hence better than any of the other more unabashedly ideological programs currently being advocated. (Hrdy 1986, as cited in Segerstråle 1992:227–228)

One does not have to identify oneself as a liberal feminist to defend the merits of standard science. Historian of science Londa L. Schiebinger (1999), for instance, who writes approvingly of postmodernist feminism, acknowledges that in many cases the feminist cause has been advanced through the use of standard methods of scholarship. Schiebinger adheres to scientific epistemology while at the same time urging that deep structural changes within the culture of science are needed if we want women to thrive in it. Since science at its origins was structured to exclude women, adding them to the process will not suffice, she argues.

Defenders of standpoint theory and postmodernism—and, hence, critics of traditional scientific epistemology—usually adhere to one of the more radical feminist traditions.

Standpoint Theory: The Objectivity of Science Questioned

Standpoint theory holds that the knowledge claims of the oppressed are truer to the world and hence more objective than the traditional scientific claims they want to replace. Many socialist feminists, black feminists, lesbian feminists, and radical feminists defend this kind of epistemology. Standpoint theorists claim that women's experiences and their traditional caring labor provide a foundation for an epistemology that is less dominating and more interactive and holistic. Their research is aimed at demonstrating how interests fashion knowledge, at uncovering supposedly sexist terminology, and at articulating alternative approaches and theories.[1] Rather than being antiscientific, standpoint theorists are science revolutionaries. Is it a potentially fruitful revolution?

As an example of feminist critiques of supposedly masculinist language, I take "The Importance of Feminist Critique for Contemporary Cell Biology" (1988), an article written by nine coauthors calling themselves The Biology and Gender Study Group.[2] The authors write that "this paper focuses on what feminist critique can do to strengthen biology. What emerges is that gender biases do inform several areas of modern biology and that these biases have been detrimental to the discipline" (Beldecos et al. 1988:173).

The Group describes how biological descriptions of the interactions between sperm and egg have frequently been modeled on various courtship behaviors.

While this may seem like a clear example of "the mask of (gender-laden) theory," influencing the resulting body of knowledge, this impression wanes, however, when we take a closer look at the examples given. It appears that, while indeed some of the language used is not emotionally neutral, there is no proof at all that "these biases have been detrimental to the discipline," as the authors assert. One passage comes from a book for expectant mothers[3]:

> Spermatozoa . . . swarm up through the uterine cavity and into the Fallopian tube. *There they lie in wait for the ovum.* As soon as the ovum comes near the *army of spermatozoa*, the latter, as if they *were tiny bits of steel drawn by a powerful magnet, fly at the ovum.* One *penetrates*, but only one. . . . *As soon as the one enters, the door is shut on other suitors.* Now, as if *electrified*, all the particles of the ovum (now fused with the sperm) exhibit vigorous agitation. (Russell 1977, as cited in Beldecos et al. 1988:175–176, emphasis added by Beldecos et al.)

The interpretation given by The Biology and Gender Study Group seems rather far-fetched:

> In one image we see the fertilization as a kind of martial gang-rape, the members of the masculine army lying in wait for the passive egg. In another image, the egg is a whore, attracting the soldiers like a magnet, the classical seduction image and rationale for rape. The egg obviously wanted it. Yet, once *penetrated*, the egg becomes the virtuous lady, closing its door to the other *suitors*. Only then is the egg, because it has fused with a sperm, rescued from dormancy and becomes active. The fertilizing sperm is a hero who survives while others perish, a soldier, a shard of steel, a succesful suitor, and the cause of movement in the egg. The ovum is a passive victim, a whore and finally, a proper lady whose fulfillment is attained. (Beldecos et al. 1988:176, emphasis in original)

As biologist Paul Gross and mathematician Norman Levitt (1994) note with regard to this passage, this is a serious case of overinterpretation. Nowhere in the literature of science will the egg be called something like a "proper lady" or a "whore." The same is true of the other examples given by the Group: The language of the quotation is far more moderate than its paraphrase, certainly in passages that they derive from scientific literature.

Surely there is nothing wrong with criticizing the use of inappropriate metaphor. Nowhere, however, do the authors succeed in substantiating their initial claim that this kind of gender bias has been detrimental to biology. Their assertion that in traditional biology the egg is still seen as passive while sperm are active is false. According to Gross and Levitt, there is now a vast and serious science of what the egg does—actively—relative to the sperm, a science that "has emerged over the past thirty years, independently of feminism or any other kind of cultural criticism" (1994:122).

Feminist language criticism has extended itself into the domain of evolutionary theorizing, demanding, for example, nonanthropomorphic language for describing animal behavior, such as 'forced copulation' instead of 'rape' (Blackman 1985; Fausto-Sterling 1992; French 1992; Harding 1985; Tobach and Sunday 1985). While this may often be a legitimate demand, it is doubtable whether changing word use has ever resulted in a reshaping of the actual body of knowledge. The implicit feminist assumption is that our language and its categories shape our thoughts and worldview. We cannot think beyond the literal meaning of concepts, and what our language does not classify, we do not see or do not attend to. Although it does not seem that far-fetched to think that language use may sometimes constrict thought, extreme versions of this view have been refuted (Brown 1991). As cognitive psychologist Steven Pinker explains in *The Blank Slate* (2002), virtually all cognitive scientists and linguists agree that language is neither a prisonhouse of thought nor a prerequisite for it. The fundamental categories of thought, such as objects, space, cause and effect, number, probability, and agency, are present in the minds of infants and nonhuman primates, although they have no language. Moreover, we do not store our knowledge in the exact words that we read or heard, but remember the *gist* of it; we have a 'semantic memory'. In cognitive science this memory is modeled as a web of logical propositions, images, motor programs, strings of sounds, and other data structures connected in the brain.

A further indication that thought does not equal language is that we don't always succeed in expressing what we really want to say. We have difficulty finding words, and often find that they do not convey what we were thinking. Because of this, and because the world is constantly changing, languages are changing as well. People coin new words and phrases to fill the gaps between thought and language. Finally, in order to understand a text we use of a lot of implicit knowledge about the world and about other people. Without realizing it, we are filling in countless unsaid steps, interpreting ambiguities, and inferring intentions. Processes of thought are going on that obviously stretch far beyond language (Pinker 2002).

Feminist critics, it seems, overestimate the power of words. There is nothing wrong in sensitizing people to inappropriate metaphor, but as a serious critique of science this search for sexist terminology turns out to be rather poor.

What about standpoint theory in general? Hilary Rose, a leading standpoint critic of science, argues that science is "an ideology having a specific historical development within the making of capitalism" (1983:274). According to her, the bourgeois and male character of science leads to highly abstract and depersonalized knowledge. She links abstraction to the alienated labor of the capitalist mode of production and to the sexual division of labor, which keeps women from entering the paid work force. In order to construct a new science and a new technology, the input of women's caring experiences is needed, thus adding "heart" to "brain" and "hand." Rose thereby stresses the "social genesis of women's caring skills" (1983:276).

Rose's analysis is a typical example of standpoint theory. Its practitioners want to reinvent science. Discarding most of 'male' logic, rationality, and abstraction, they stress subjectivity, concreteness, relatedness, and inclusiveness. The goal is to develop 'female sciences', such as female logic, mathematics, physics, astronomy, and philosophy.

Since the project of reinventing science sounds quite spectacular, my evaluation of it takes more time than that of the empiricist position. We will have a look at its problems and its results respectively.

Standpoint theory sees itself confronted with an inherent contradiction. If standard science is in essence the product of the ruling classes, then standpoint theorist scientists cannot but engage themselves in a capitalist, patriarchal enterprise in trying to correct it. Hilary Rose recognizes this problem when she says that "[t]he trouble with science and technology from a feminist perspective is that they are integral not only to a system of capitalist domination but also to one of patriarchal domination" (1983:266). To try to discuss science under both these systems is peculiarly difficult, she says, adding that historically, it has been women outside science, such as the novelist and essayist Virginia Woolf, who have dared to speak of science as male, as part of a phallocentric culture.

For Rose, the explanation of the dearth of female scientists who attack the underlying conceptual basis of science seems to go without saying: "For women inside science, protest has been much more difficult. Numbers are few and developing the network among isolated women is intractable work" (Rose 1983:266–267).

There might, however, be another explanation of the fact that so few female scientists appear to regard science as a patriarchal construction: Since they are thoroughly acquainted with scientific methodology, they simply *know* that it is not inherently male. Many feminist standpoint theorists are familiar with the practice of science only through the work of sociologists of science, and in the field of the latter, social constructionism reigns partly because many sociologists of science are not acquainted with scientific methodology either. As Susan Haack argues, good sociology of science, which acknowledges the relevance of evidential considerations, requires some grasp of scientific theory and evidence, while bad sociology of science does not. She adds that "[a] cynic might suspect that this partially explains why there is so much bad sociology of science about" (1996b:260).

Yet Gross and Levitt are probably not being cynical when they state that common to almost all critiques of standard science is "a failure to grapple seriously with the detailed content of the scientific ideas they propose to contest" (1994:235). This is because the critics are not familiar enough with the sciences they criticize. They simply sail so wide of the mark, and have so little to say about the actual ideas with which scientists contend, that the latter generally ignore these critiques. To put the matter bluntly: Only for people who do not know what they are talking about does the radical attack on standard science sound reasonable—apart from a small group of ideologically motivated scientists. One simply cannot make profound criticisms of a technical subject of which one is entirely ignorant.

A second problem standpoint theory has to contend with is cognitive relativism. Sandra Harding (1986) wonders whether there should not also be Native American, African, and Asian sciences and epistemologies, based on the distinctive historical and social experience of these peoples. She thereby points to "the fatal complication for this way of thinking—the facts that one-half of these peoples are women and that most women are not Western" (1986:297). Indeed, on what grounds would feminist sciences and epistemologies be superior to those of other oppressed groups? Harding suggests that we might regard them all as valuable in their own right, not as one being better than another. Thus standpoint theory easily slips into complete relativism, with one difference: The knowledge claims of "the oppressed" are regarded as truer than those of "the oppressers."

Although science has its roots in medieval Arabic culture, it is undeniable that the modern scientific method originated in Western Europe and not anywhere else. It might, however, be wrong to speak of *Western* science, because there are no indices that other sciences might exist:

> [T]o a certain extent, genuine science emerged as much in spite of as because of Western culture. It was vigorously opposed, after all, by the foundational institution of Western civilization, the Roman Catholic Church, and a continuous line of Western thinkers . . . has always stridently opposed scientific methodology if not rationality itself. Far from being an attribute of Western culture, scientific thought transcends cultural context. All scientific claims, of course, are shot through with cultural expectations, but science itself is a self-correcting process that over time weeds out the culturally contingent for the universal. It was precisely because of its universalism that modern science diffused so rapidly across the globe, creating a contemporary scientific project that is remarkably cosmopolitan. (Lewis 1996:216)

Moreover, as Lewis points out, Western Europe is wrongly represented as an isolated entity. Well into the modern period it was just one of a series of linked

Afro-Eurasian civilizations that had historically moved more or less together in economic and technological evolution and that even supported similar ideological structures. Nor can Western Europe lay any special claim to rationality. Historically, reason and unreason have coexisted in every civilization, in the West as well as in the East.

If one wants to contrast traditional Far East medicine with Western medicine, he continues, then

> one must examine the *traditional* medical practices of western Europe—in other words, the humoral theory of disease, associated with such "curative" practices as massive bloodletting. Such medical "science" was, after all, standard throughout the West until well into the nineteenth century. It would eventually yield to modern medicine, not least because its own methods were often deadly. If "Eastern medicine" survives to this day as an alternative to modern healing—in California no less than in Japan—it is largely because its "cures" are at worst rather innocuous. (Lewis 1996:216, emphasis in original)

Since the end of the eighteenth century other ethnicities have increasingly come to be represented in science, such as Jews, Indians, Arabs, Pakistanis, and Chinese. As individuals, many of them have made contributions of the first rank and of enormous influence, but they did not invent a new multicultural science. Indeed, it is unclear how they could have, even if they had wanted to. Take Goethe's battle against Newton over color and light. For Goethe, light was a fundamental entity of nature and must not be decomposed. He waged a war against abstraction and the use of experiments, insisting instead upon a closeness to and an identification with nature. This substitution of intuition for logic, analysis, and abstraction, a characteristic of romantic natural philosophy, was a root cause of its failure to produce useful science (Gross and Levitt 1994).

The supposedly nondualistic worldview of the East did not prevent environmental degradation. Similarly, traditional, non-Western worldviews have never protected women from discrimination; on the contrary. We might argue that, instead of being bad for women, properly done science combined with rational, open discussion of its results is a liberating force, not only for women, but also for all oppressed groups.

Many liberal feminists consider the rejection of rationality as contrary to women's purposes.[4] Regarding "white, male thought" Cathy Young claims that "it would be ludicrous for women to deny themselves that heritage" (1999:38). Philosopher of science Noretta Koertge puts it more sharply:

> If it really could be shown that patriarchal thinking not only played a crucial role in the Scientific Revolution but is also necessary for carrying out scientific en-

> quiry as we know it, that would constitute the strongest argument for patriarchy that I can think of! I continue to believe that science—even white, upperclass, male-dominated science—is one of the most important allies of oppressed people. (1996b:413)

Scientific Contributions of Standpoint Theory

Have standpoint theorists, with their epistemology of holism, harmony, and care, made any important contributions to science? Many feminists see the work of biologist Barbara McClintock as exemplary (Donovan 2000; Fox Keller 1982; Rosser 1992). In the 1940s McClintock discovered that pieces of DNA can "jump around" on the chromosomes. After having been ignored by the scientific establisment for decades, her discovery was finally confirmed and in 1983 she was awarded the Nobel Prize (Ridley 1999).

As Evelyn Fox Keller (1982) and Sue Rosser (1992) describe it, McClintock's discovery did not fit within the then predominant view of genes as immovable beads on a string and of DNA as the "master molecule," encoding and transmitting all instructions for the unfolding of a living cell. Hers was a more interactionist model. Fox Keller has interpreted McClintock's insistence on the importance of "letting the material speak to you" and of having "a feeling for the organism" as particularly feminine (1982:243). For Fox Keller, McClintock's research is definitely an example of what a different science, "a science less restrained by the impulse to dominate," might look like (1982:245).

McClintock herself has, however, never accepted Fox Keller's interpretation of her work as representing a woman's perspective. To her, science was not a matter of gender, but a place where, ideally, the matter of gender drops away (Hoff Sommers 1994; Radcliffe Richards 1995).

Gross and Levitt (1994) share McClintock's view. They point out that her closeness to the experimental material is characteristic of some scientists and less so of others. Moreover, there is no lack of abstraction in her work; it is solidly grounded in the abstractions of formal genetics. McClintock simply saw things that others didn't see. They argue further that good experimentalists *must* be close to the experimental object in order to make effective designs; otherwise they are unable to identify and exclude intervening variables.

As to the idea of DNA as "master molecule," Gross and Levitt write that it is not, even among the most radically reductionist biologists, a literal one: Only secondary writers think *ex DNA omnia*. It is just a shorthand for "initial information source," nothing more, and it carries no implication of dominance. Interactive relationships have been a central concern of cell and molecular biology for more than forty years.

If McClintock's work does not testify to a feminine way of knowing, then perhaps more recent feminist research does? In trying to assess this matter, it is important to keep in mind the difference between 'female contributions to science' and 'female science'. It is undeniable that because of the underrepresentation of women in the social and life sciences, the actual practice of science has sometimes failed or still fails to live up to its norms. These norms comprise logical clarity, logical consistency, generality of principles, and empirical validity.

When, for example, medical or psychological theories are not adequately tested on female subjects, the theories lack generality of principles and empirical validity. Science presents other genuine issues of concern. Laboratories can be very unwelcoming to women, for instance. Women scientists sometimes find that their work is not taken as seriously as that of comparable male colleagues and that they may not be included in informal communication networks. A Swedish study (Wennerås and Wold 1997) showed that women scientists need to publish two and a half times more papers to rank equally with male colleagues. Other studies revealed that a paper attributed to a man is rated more favorably by both sexes than when the same identical paper is attributed to a woman (Schiebinger 1999). Thus women can still change a lot in the world of science, apart from bringing in new heuristically valuable points of view and providing additional sources of scrutiny.

This is not the same, however, as developing a *female* science, with a new methodology, new rules of evidence, and another interpretation of objectivity. And the fact is: Feminists do not seem to have been able to develop this kind of science yet, and it is doubtful whether they ever will. In her description of the various strands of feminism, Sue Rosser (1997) offers examples of the scientific contributions within each tradition. When we take a closer look, however, we see that her examples are the same for *all* feminist traditions that epistemologically adhere to empiricism or standpoint theory: flawed research within anthropology and primatology brought to light by regular women scientists such as Sally Slocum, Sarah Hrdy, and Jane Goodall.[5] There is no sign of a female way of knowing; this is just standard science, albeit that as women those scientists were more focused on female subjects in their research. Rosser's examples testify only to the self-correcting nature of science.

In discussing radical feminism and its complete rejection of current scientific epistemology, Rosser confesses that a new female science has not yet been constituted:

> It is difficult to imagine what might result if women scientists congregated regularly for sustained periods in the absence of men to decide what constitutes knowledge. Moreover, if women scientists also dominated and controlled the leadership positions within the scientific establishment, the government, and academia for decades, who knows what might emerge? (1997:33–34)

Thus, following Rosser, we may conclude that as yet there is only standard science—although that is surely not what she intended to demonstrate.

Other authors concur. Mathematician Mary Beth Ruskai does not know of any cases in mathematics in which someone has alleged that a theorem is wrong because a woman proved it. If a result is right, it is right. She continues:

> In some of the gender-difference literature the arguments seem to me particularly flawed because they define feminine in terms of certain well-known, common characteristics. They also define science in terms of other characteristics (sometimes of questionable validity, such as noncreative and nonintuitive) that they consider masculine. If one points out that there are many examples of women who do not fit this stereotype, the response is that the paradigm is cultural rather than biological, so that the women who do not fit are masculine and the men who do fit are feminine. To a mathematician this sounds suspiciously like "assuming what you're trying to prove." (1996:440)

An example of feminist algebra is described from a critical perspective by Gross and Levitt (1994). It concerns, however, again a case of "unobjectionable," but "futile" linguistic alterations (1994:114). The authors of their sample disapprove, for example, of a particular problem in which a girl and her boyfriend run toward each other because it portrays a heterosexual involvement. Or they object to a problem about a contractor and the contractor's workers (sex undeclared), because they assume that the student will envision the workers as male.

According to Gross and Levitt, "[f]eminist cultural analysis has not yet identified any heretofore undetected flaws in the logic, or the predictive powers, or the applicability of mathematics, physics, chemistry, or—much complaining to the contrary notwithstanding—biology" (1994:112). Whether their mentioning of biology in this context stands up to scrutiny depends, I think, on the exact interpretation one prefers to give to the terms "predictive powers" and "applicability." Does the long neglect of female-female relationships in ethology constitute a flaw in the predictive powers and applicability of ethology? One could say that, indeed, it did for a long time, until this flaw was corrected by female ethologists. One could also say that the fact that female ethologists found this flaw testifies to the built-in self-scrutiny of the scientific enterprise and thus to its ultimate applicability. The same is true of research in the health sciences and other life sciences.

Psychology does not seem to have been altered by 'female' science either. Psychologist Lynne Segal wonders about it that within psychology feminism has made a comparatively small impact. Feminist critique of psychology's statistical methods has remained marginal, she observes, "despite the remarkable shift in the numbers of women entering the discipline: 400 per cent *more* women than men studying psychology today, compared with the 50 per cent *less* women than men back in 1969" (1999:151, emphasis in original). Thus, although no doubt women psychologists will have brought new perspectives and accents to the discipline, they have not changed its methodology. The use of logic, statistics, and abstraction remains as valuable as ever.

Feminist ethologists Donna Holmes and Christine Hitchcock (1997) have examined the research preferences of male and female scientists in their field. As they say, the science of animal behavior is a field where women are participating comparatively equally as professionals, at least relative to biology as a whole.

Holmes and Hitchcock surveyed almost 2,600 abstracts from presentations of national meetings of the Animal Behavior Society from 1981 to 1990. They found that both sexes study mammals most often. Women are, however, still more likely than men to study them. When all animal taxa were analyzed together, their data failed to support the hypothesis that women are more likely than men to focus on female animals or on young. When they investigated primate studies only, however, they found a marked tendency to focus on animals of the investigator's own sex. This may not surprise us, since primates are "animals with which we all presumably have a heightened capacity for empathy or intimacy, or a 'feeling for the organism'" (Holmes and Hitchcock 1997:196). Women were in general more likely than men to focus on some aspect of social behavior, especially parental care. Social behavior was, however, one of the topics investigators of both sexes mentioned most often. There was no evidence that men were more likely than women to focus on aggression or male combat, but researchers who studied sexual selection, regardless of sex, were much more likely to focus primarily on the sexual behavior and combat of males.

Thus Holmes and Hitchcock find that there are (small) sex differences in choice of research topics by ethologists, and that these differences will grow when research focuses on animals one can more easily identify with: Male primatologists tend to focus on male apes and female primatologists on female apes. A heightened "feeling for the organism" thus seems to be more closely related to the kind of organism studied than to the investigator's sex. By extension, we can expect identification with one's own sex to be still stronger in the study of human subjects.

This testifies again to the need of female researchers *in disciplines in which sex is itself intrinsic to the subject being studied*, that is, in the life sciences and the social sciences. In disciplines such as mathematics and physics, the sex of the investigator is not expected to make any difference. There is no indication, however, that men and women do science in a qualitatively different way (a question that Holmes and Hitchcock did not directly address; they think that to investigate this may prove an impossible task).

The data thus seem to indicate that what some feminists consider 'female' science is just standard science performed by female researchers. At the end of this chapter we see that this observation was to be expected on strictly logical grounds.

Postmodernism: The Whole Project of Science Questioned

The postmodernist line of thought has already been sketched in the introduction. As Harding writes, a postmodernist view of science holds that standpoint theory is still too firmly rooted in "distinctively masculine modes of being in the world" (1986:294). It implies the complete rejection of rationality and of dispassionate objectivity and thus seems to challenge the legitimacy of trying to describe the world from a feminist perspective.

Postmodernist critics of science are resistant to any generalizing theory, fearing that it necessarily represses or ignores contradictory evidence. Adherents of queer theory, radical feminism, some strands of psychoanalytic feminism, and postmodernist feminism can often be found defending this epistemology. Many ecofeminists share its complete rejection of 'male' rationality.

The line between standpoint theory and postmodernism is vague, however. As indicated above, a postmodernist kind of epistemology can sometimes be found within the writings of standpoint theorists as well.

The contradiction at the heart of postmodernist epistemology is this: If all scientific statements are historical fictions made facts through the exercise of power, then how can feminists produce knowledge that is more *true*? Postmodernist feminists do not seem to find the answer. They treat the problem in a highly abstract way, such as by saying that they have a dream "not of a common language, but of a powerful infidel heteroglossia" (Haraway 1991d:181) and by talking about "situated and embodied knowledges" that sustain "the possibility of webs of connections called solidarity in politics and shared conversations in epistemology" (Haraway 1991f:191). Radical feminists sometimes use consciousness-raising groups, in which they meet to examine their personal experiences, to determine what counts as knowledge.

Philosopher of science Mario Bunge notes about cognitive constructionists that it would "go against the grain of their school to propose its own clear-cut criteria of scientificity" (1996:106). But, he wonders, how is it possible to discuss rationally the scientific status of an idea or practice otherwise than in the light of *some* definition of scientificity? He thereby touches upon the logical inconsistency within the phrase "feminist epistemology" that will be elaborated at the end of this chapter.

Should Science Be Politically Progressive?

There exists a deep divide between, on the one end of the epistemological continuum, those feminists who believe in value-free science and who consider scientific rationality a source of empowerment for women, and, situated near the other end, those who explicitly reject standard science and rationality as inherently patriarchal. The second attitude leads to the demand that science should be guided by political commitments.

Although sounding rather outrageous at first hearing, this demand follows quite logically from the social constructionist line of thought. Since constructionists believe that science is always disguised ideology (to a greater or lesser extent, depending on the theorist), they want to infuse it with the "correct" political and ideological values. On this view, science not only is but also should be politics by other means. The aim no longer is value-free science.

As an example I quote feminist biologist Anne Fausto-Sterling from her influential book *Myths of Gender*, an attack on biological theories about sex differences. She describes a scientist's standard of proof as "a standard dictated in turn by political beliefs" (1992:11), adding:

> I impose the highest standards of proof, for example, on claims about biological inequality, my high standards stemming directly from my philosophical and political beliefs in equality. . . . This book is a scientific statement *and* a political statement. It could not be otherwise. Where I differ from some of those I take to task is in not denying my politics. (11–12, emphasis in original)

When reading the book, however, one gets the impression that *no* scientific research will ever meet her standards of proof, simply because she a priori assumes that almost all differences between men and women are socially constructed. Since her political beliefs determine which theories can be valid, she will never accept theories that are not to her liking.

As I pointed out in this chapter and the previous one, there are good reasons for believing that social constructionist epistemology is built on quicksand. In that case, its argumentation for making science explicitly political becomes invalid. If science, as a collective and cumulative enterprise, is (ideally) value neutral, there is no logical necessity to let it be driven by "better" values.

But on the face of it, the suggestion to harness science to the task of making our society more just and more politically progressive has an attractive humanistic ring to it, as philosopher of science Noretta Koertge (1996a) admits. Who can be against justice and political progress? And since the quest for pure scientific truth is already restricted by ethical prohibitions on harming experimental subjects and mitigated by a responsiveness to social needs, why not make an opening for deliberate ideological intervention?

How would the constructionist proposal impact on the process of scientific inquiry? In the traditional account, the context of application of scientific results lies outside the domain of science proper, and it is here as well as in the choice of funding priorities that social values and policies take their rightful place. Now everyone would agree that extreme care should be taken in applying scientific results that are still preliminary or could easily cause harm, Koertge argues. As she says, however, the constructionists go much further, asserting that some research is so politically dangerous that it should never be done. Hence, they try to suppress publications or research conferences on possible biological correlates of crime, intelligence, or sexual orientation. They thus introduce political or ideological considerations directly into the heart of the context of justification of research. So we should, for example, study the social construction of sexism, not sex differences, and analyze the origins of homophobia instead of the etiology of homosexuality.

Such an ideological filter could seriously reduce the number of topics that can be investigated and damage the ethos of free inquiry so treasured in science. That is the first reason why the constructionist proposal should not be defended. The second reason is yet more important. Whereas censure of specific research topics would not in itself distort the content of scientific findings, such a distortion would happen if ideology were to affect the process of *hypothesis formation*, something that is defended by feminist constructionists. As the quotation of Helen Longino at the beginning of this chapter makes clear, they want to subordinate science to their political aims. On their agenda, the scientific acceptability of a claim would depend on its perceived political suitability.

History should have taught us by now the futility of such a strategy, Koertge (1996a) argues:

> [T]he act of banning Copernicanism hurt the Catholic Church much more than heliocentrism could ever have done. Lysenkoism[6] did nothing to help peasants improve their agricultural practices while it delayed the adoption of hybrids developed in the West. Women will not benefit in the long run from attempts to block the study of biological differences. (Koertge 1996a:271)

Thus her conclusion is that a constructionist epistemology can only be detrimental to women. Not only is it a waste of time and energy to try to block standard science, it also keeps women from participating in science and thus from gaining power. As Lopreato and Crippen argue, "[w]hen science and politics are mixed, both activities may suffer" (1999:69).

Does Feminist Science Exist?

In her cogently argued article "Why Feminist Epistemology Isn't (and the Implications for Feminist Jurisprudence)" (1995), philosopher Janet Radcliffe Richards investigates what 'feminist knowledge' and 'a feminist epistemology' do actually mean. Investigating the foundations of science, epistemology, logic, or ethics is a perfectly respectable philosophical activity, in which feminists seem as entitled to join as anyone else, she says. In a long line of argument, she subsequently demonstrates, however, that feminism can provide no justification for holding one theory rather than another. Ultimately each scientific theory, whether it calls itself feminist or not, is evaluated by the same traditional epistemological standards that *everyone* uses, although some may not recognize that they use them. It is simply logically impossible to reject standard epistemology.

Radcliffe Richards starts her argument by showing that, although feminist epistemology is supposed to be completely at odds with traditional epistemology, there is some common ground. To be a feminist obviously implies, after all, having a view about the way things *are*, otherwise one could not think there was anything wrong with the position and treatment of women. These beliefs about the way things are, are supported by empirical, often scientific, investigation. This means that one also has beliefs about epistemology and the scientific method, since one's conclusions about what the world is like depended on assumptions about how these things can be found out.

Similar points apply to questions of value. In order to complain about the position of women, one must be appealing to moral standards of right and wrong. The belief in such standards in turn implies beliefs about metaethics, about the way one reaches ethical conclusions.

Thus feminism necessarily starts from the appeal to traditional standards of ethics, epistemology, and science and from the use of traditional logical inference. It was by appeal to these standards that the position of women was first found to be wrong, for feminism as a movement as for individual feminists. At this stage the feminist's beliefs are still not feminist in the sense of there being any reason for a feminist to hold them that a nonfeminist—someone who has not yet recognized that women have grounds for complaint—does not have. The feminist can, for example, come to the conclusion that traditional ideas about women cannot be justified by traditional epistemological standards, and she can demonstrate this to any impartial investigator who shares her basic standards.

Here an important point in Radcliffe Richards's argument follows, "simple and obvious once seen," but nonetheless "widely overlooked":

> You may change your views because of feminism in the sense that you would not otherwise have embarked on the inquiries that led to those changes, but that does not mean you change them because of feminism in the sense that feminism

> provides the justification for the change. . . . Feminism may set investigators on the track of new discoveries, and the feminist program may expand as a result of them, but that provides no . . . reason to count the new beliefs themselves as feminist. (1995:373–374, emphasis in original)

The new beliefs are thus the result of the application of standards of science and rationality that are not themselves feminist. The question now is: How might the feminist be *led by her feminism* to reject these traditional ideas about reason and science and eventually adopt radically different ones?

Of course, by its very nature, feminism must challenge some received ideas. As Radcliffe Richards explains, however, standards of knowledge come in hierarchies, and many changes can be made at intermediate levels long before the fundamentals of rationality and epistemology are even approached. It is in particular essential to recognize that changes in beliefs about what the world is like and how it works always amount to potential changes in standards, because these beliefs are automatically used as the basis for assessing other, less well-established beliefs. In forensic medicine, for example, we decide guilt by reference to our fundamental beliefs about blood groups or DNA, whereas once, long before forensic medicine existed, we might have decided it by whether the accused sank or floated in water. Thus it is likely that feminism leads eventually to extensive changes of standards at this intermediate level. If the idea is challenged that women's nature allows them to find happiness only through marriage, for instance, this will lead to a fundamentally different approach to the assessment of women who feel unhappy about their lives.

Another example: Suppose feminists are right in claiming that women have ways of knowing that have been disregarded because they are different from paradigmatic scientific procedure. If feminist-inspired investigation eventually showed not only that such female techniques exist, but also that they were just as effective, by ordinary criteria for scientific success, as the paradigmatic ones, that would support feminist demands for changes in the standards for the assessment of knowledge claims. It would do so, however, only on an *intermediate* level. There would be no reason for changes in epistemology or fundamental attitudes to science. Quite the contrary, as Radcliffe Richards asserts: Whether or not these female ways of knowing would be accepted as reliable *depends*, once again, on the acceptance of more fundamental standards of science and epistemology. If they were shown to be reliable, they would then be included into traditional scientific procedure and become one of the standard ways of investigating the world.

To show that more fundamental changes were needed in the criteria for scientific success, it would be necessary to imagine women's being *unsuccessful* by current standards—having theories that tests kept showing were getting nowhere, making predictions that were usually unfulfilled, and so on—and then saying that scientific standards should be changed to count *this* as good science. As she says, it is difficult to imagine either what such standards would be, or that any feminist would recommend them. She continues with a claim similar to the one made by Susan Haack (1996a):

> [I]t would be a serious mistake to respond by saying that if these [knowledge claims] were what traditional epistemology counted as knowledge, there must be something wrong with traditional epistemology. This would not only be much too precipitate; it would also give far too much credit to patriarchal man. It would by implication concede that whatever he had claimed as knowledge must, by traditional standards, really be knowledge, to be dealt with only by complicated revolutionary epistemologies through which it could be shrunk into mere phallocentric knowledge, or otherwise emasculated. It is usually much simpler—and, you would think, much more feminist—to start with the assumption that what has been *claimed or accepted* as knowledge may not be knowledge of any kind, even phallocentric, but, by patriarchal man's very own epistemological standard, plain, ordinary (perhaps patriarchal) *mistakes.* (Radcliffe Richards 1995:377, emphasis in original)

Thus the fact that claimed proofs can be mistaken does not even begin to show that there is something fundamentally flawed about traditional epistemological ideas of propositional knowledge. It is Radcliffe Richards's suspicion that a great deal of the impulse toward feminist involvement in epistemology and philosophy of science arises from the blurring of this distinction between the epistemological standards themselves and a wrong application of them.

An indication that confusions of level may be a source of problems is the typical shift in feminist writings from the discussion of a particular aspect of science or its applications to a general conclusion about science in general, without systematic discussion of how the different levels relate to each other. She describes how she has "often been struck in practice, for instance, by the way feminists who are (quite reasonably) angry about the male takeover of obstretics describe the insensitive use of gadgets as 'subjecting women to male science,' then go on to take this as indicating some kind of global, woman-oppressing maleness of every aspect of the scientific enterprise" (1995:378).

Radcliffe Richards's essential point is this: One cannot logically leave traditional epistemology behind *and* keep earlier conclusions about the position and treatment of women that were actually based on the rejected epistemology. Criteria for proper treatment and assessment need to be in place before it can be said that the present state of things is falling short of them. Feminist theories are *at all stages* dependent on more fundamental ideas that have nothing to do with feminism. No beliefs about matters of fact, and no theories of ethics, epistemology, or science, can be required by feminism, because feminist conclusions depend on them.

Even the argument that we must adopt the epistemological ideas claimed as feminist because the progress of women depends on it presupposes the idea that epistemology is a sort of thing that is logically secondary to ethics, which is itself an epistemological theory. Moreover, if we can tell what is going to benefit women, we must think we know something about the way the world works and therefore must presuppose an epistemology other than the one we are supposed to be defending.

Radcliffe Richards's conclusion is clear: There can be no *feminist reasons* for adopting beliefs of any kind. Feminism can provide no justification for holding one theory rather than another. If a theory that goes by the name of feminism turns out to be good science, it just becomes standard science. *The fact that some belief is held by a feminist does not make it a feminist belief.* Feminism may prompt women to raise particular questions, but it cannot be the determinant of the answers. The worthwhileness of ideas must be assessed in complete independence of their claimed connection with feminism.

If one really wants to fight the oppression of women, there is only one way to go, according to Radcliffe Richards:

> Once you have reached such conclusions as that traditional claims about the nature and position of women have always been inadequately supported by the evidence and are often mistaken, that women have had various direct and subtle obstacles placed in the way of their acquisition of knowledge, and that their abilities have never been impartially assessed or their achievements acknowledged, that shows the direction your feminist politics must take. You must therefore be positively opposed to any movement that encourages women to disdain as patriarchal the very ideas in philosophy and science you have been trying to ensure they are given full opportunity to learn, or that demands recognition as special ways of knowing for the contrived ignorance you have been struggling to remedy. . . . And if such a movement calls itself feminist, that only provides a reason for redoubling your efforts to thwart it. (1995:392)

The qualification 'feminist' with regard to science and epistemology is thus unnecessary or even misleading. Nothing is lost by approaching the issues of feminism in this way, Radcliffe Richards argues, since it

> leaves completely open all questions about what theories, standards and methods should be accepted, how different the sexes actually are, how much injustice to women still remains, and everything else. It does not in the least imply that the approaches now claimed as feminist should *not* be adopted; all it does is undermine the spurious status the feminist label has given them. (1995:400, emphasis in original)

With the issue of feminist knowledge and feminist science thus clarified (or so I hope), we can now proceed to the history of evolutionary theorizing. Whenever I refer to 'feminist criticisms' of Darwinian theory, this will mean 'formulated by feminists'. We see that, whenever these criticisms turned out to be well-founded (as assessed by traditional epistemological and scientific standards), they were sooner or later incorporated within evolutionary theorizing—although often later than sooner.

The history of Darwinism testifies to the earlier-mentioned argument Radcliffe Richards developed: The only productive way for women to correct male bias in science is not by dismissing the scientific endeavor as 'patriarchal', but by engaging in it themselves.

Notes

1. For example, Beldecos et al. 1988; Bleier 1985; Dines, Jensen, and Russo 1998; Fausto-Sterling 1992, 1997, 2000a, 2000b; Fox Keller 1982; Harding 1986; Hubbard 1990; Rose 1983; Rosser 1992.

2. The authors are Athena Beldecos, Sarah Bailey, Scott Gilbert, Karen Hicks, Lori Kenschaft, Nancy Niemczyk, Rebecca Rosenberg, Stephanie Schaertel, and Andrew Wedel.

3. K. P. Russell (1977), *Eastman's Expectant Motherhood.*

4. Campbell 2002; Hoff Sommers 1994, 2000; Koertge 1996a, 1996b; Nanda 1996; Patai 1998; Radcliffe Richards 1995; Roiphe 1993.

5. Jane Goodall had no scientific background when she left for Tanzania to study chimpanzees in the wild (Goodall 1999). Her work is now, though, generally acknowledged as being of primary scientific importance.

6. The "Lysenko affair" refers to the fact that from about 1935 to 1965 the development of genetics in the Soviet Union was halted. Mendelian genetics was replaced with a doctrine about the inheritance of acquired characteristics championed by Trofim Lysenko, a political figure put in charge of attempts to improve agriculture (Segerstråle 2000; see chapter 4 for information about acquired characteristics).

Sociologist Ullica Segerstråle adds this interesting remark on Lysenkoism:

> The Lysenko affair has become something of an exemplar referred to by academics to emphasize the importance of freedom of science, or as an example of what happens when scientific objectivity is abandoned for political reasons. It seems that the Lysenko affair is as routinely invoked by scientific traditionalists as the notorious sterilization laws or Immigration Act are invoked by radical scientists. Ironically, both sides use their chosen precedents in order to demonstrate the *same* thing: the dangers of politically influenced science. (2000:228, emphasis in original)

It is, though, not exactly the same thing they want to demonstrate. Whereas scientific traditionalists want to show what happens when you let scientific results be influenced by politics, radical scientists maintain that science is *always* politically influenced and that, therefore, we better be selective in choosing our topics of research.

3

The Sexes since Darwin

> [T]he sexes in each species of beings compared upon the same plane, from the lowest to the highest, are always true equivalents—equals but not identicals in development and in relative amounts of all normal force. This is an hypothesis which must be decided upon the simple basis of fact.
>
> Blackwell 1875:11

As we see in this chapter, there can be little doubt that Charles Darwin (1809–1882) let his theorizing on the sexes be constrained by his Victorian worldview. Yet it might still be argued that his proposition of female choice in sexual selection was revolutionary at the time. When reading feminist accounts of Darwinian theory, however, one gets the impression that no feminist wants to acknowledge this element of female agency within Darwin's writings (with 'feminist' I mean 'non-Darwinian feminist'; in referring to Darwinian feminists I will always explicitly use the appellation 'Darwinian').

An example of this typical kind of selective feminist reading is Donovan (2000), who summarizes the content of Darwin's major work on sexual selection, *The Descent of Man and Selection in Relation to Sex* (1871), in one sentence, saying that the book "promulgates a doctrine of male superiority" (2000:58). Now it cannot be denied that Darwin indeed thought that "man has ultimately become superior to woman" (Darwin 1998 [1871/1874]:585). In evaluating his view of women it seems, however, unfair to focus only on the embarrassing mistakes he made. Moreover, the observation that he could not prevent Victorian values from slipping into his theorizing does not at all detract from the scientific value of evolutionary theory. As Antoinette Brown

Blackwell, the first woman to publish a critique of Darwin's and Herbert Spencer's view of the sexes, argues in the quote at the opening of the chapter, the precise relationship between the sexes should be decided upon the basis of fact.

According to Darwinian feminist Sarah Hrdy in her preface to the second edition of *The Woman That Never Evolved* (1999a), female strategies are now center stage in sociobiology, and there is an unprecedented funding for female-focused research in this field. She adds: "One of the great strengths of science, after all, is the power to correct (however belatedly) erroneous ideas" (1999a:xviii).

It took indeed more than a century for Antoinette Brown Blackwell's hypothesis to be seriously tested, a sad fact that will not exactly serve to decrease the feminist suspicion of science. It is, then, not entirely without reason that many feminists have judged an evolutionary account of the psychosexual differences between men and women to be sexist.[1] But although this negative reaction may be understandable, I am convinced, as are other Darwinian feminists, that feminism throws away a valuable tool for understanding sexism in denouncing an evolutionary approach to the human mind.

In this chapter I give an account of the history of evolutionary biology from the nineteenth century to the period just before the official birth of sociobiology in 1975. It will not be a standard survey in the sense that I write it from a Darwinian feminist perspective: focusing on the (lack of) attention to females in the historical process of evolutionary theorizing and evaluating feminist criticisms of the discipline.

The World before Darwin

Living in the twenty-first century, with evolutionary theory firmly established in most parts of the Western world, it is difficult to imagine what it must have been like to live in a pre-Darwinian world. Much of natural history was closely wedded to natural theology, which held that all species were consciously designed by God, as described in the Bible. Species were considered immutable; they had an 'essence', created by the deity. In this statical and essentialist worldview, there was no place for evolution. Although evolutionary ideas had been around since the eighteenth century, they were not very influential (Braeckman 2001).

Historically the most serious alternative to Darwinian theory has been proposed by the French biologist Jean-Baptiste Lamarck (1744–1829). His theory as it was taken up in Britain can be summed up in the phrase 'use-inheritance'.[2] The idea is that the activity of an organism shapes the organism appropriately for that activity, and that these "acquired characteristics" will then be passed on to the organism's offspring. Adaptations arise as a response to needs. The more a giraffe stretches its neck in order to reach the leaves it feeds on, the longer its neck becomes, and its offspring will then inherit this longer neck. Toward the end of the nineteenth century, the German biologist August Weismann showed the inheritance of acquired characteristics to be impossible (Cronin 1991).

Evolution by Natural Selection

By the middle of the nineteenth century, the idea of evolution gradually came to be accepted. Evolutionists were unable, however, to pin down a mechanism of change. They called in conscious design or remained stuck in vagueness. It was Darwin's great triumph that he found a mechanism for evolution that explained adaptive complexity, that is, the careful fit between an animal or plant and its environment. It also solved the puzzle of the astonishing diversity combined with fundamental similarity among groups of organisms (Cronin 1991).

Darwin called this mechanism 'natural selection', the workings of which he explained in *On the Origin of Species by Means of Natural Selection: Or the Preservation of Favoured Races in the Struggle for Life* (1859). Natural selection consists of three essential ingredients: variation, inheritance, and selection. There is always variation among organisms. Some of these variations can be inherited, others cannot. Selection can only work on the heritable ones. As a blind and nonintentional force, natural selection acts as a kind of sieve. An organism with a variation that makes it better adapted to its present environment will be more successful at surviving and/or reproducing relative to others. It will thus leave relatively more offspring. The favorable characteristic will be passed on to future generations at ever greater frequency, until it has become a characteristic of the species itself—that is, as long as the environmental circumstances do not change in such a way that the characteristic becomes detrimental to the organism. The various shapes of the beaks of the finches at the Galapagos Islands, for instance, are adaptations to the different types of flowers each kind of finch feeds on (Darwin 1964 [1859]).

The process of sexual selection, which is, simply said, not about the struggle for life but about the struggle for reproduction, is only briefly mentioned in *On the Origin.* A full elaboration of the theory would have to wait until 1871, when *The Descent of Man and Selection in Relation to Sex* was published.

Initially Darwin's theory of natural selection met with a lot of resistance, also from biologists. One correct objection was that it lacked a coherent theory of inheritance—the fusion of Darwinism with the discovery of gene inheritance would not happen until the 1930s. Another objection was that it was difficult to imagine how the early stages of the evolution of an adaptation could be useful to an organism. It is now known that partial forms can indeed offer adaptive advantages. Many people had, as was to be expected, problems with the materialism of the theory: They could not believe that the emergence of humans and other species resulted from the slow, unplanned, cumulative process of selection, instead of them being created ready-made by God (Buss 1999).

Today the modern Darwinian theory of evolution is the unifying and nearly universally accepted theory within the biological sciences (Buss 1999). Feminist critiques of it have focused mainly on the aspect of sexual selection. Some feminists, however, have tried to undermine Darwin's scientific credibility by pointing to the congruence between his theory of evolution by natural selection, with its focus on competition, and the values of upper-class Victorian England (Bleier 1985; Haraway 1991e; Hubbard 1990; Rose 2000; Rosser 1992, 2003).

As a biologist, Sue Rosser can of course not deny "the significance of the theory of natural selection for the foundations of modern biology" (1992:57). Note her choice of words: She does not positively state that she believes the theory to be *true*. Her other remark speaks for itself: "[S]cholars have suggested that natural selection, as described by Darwin, was ultimately accepted by his contemporaries because it was a paradigm laden with the values of nineteenth-century England" (1992:56). It must be said, however, that it was the notion of *evolution* that was accepted. Natural selection remained a contentious issue for a long time.

Like Rosser, Ruth Bleier suggests that Darwin's theory of evolution cannot be separated from its genesis in a capitalist society and thus cannot be neutral science:

> Historians of science have documented the intertwining, interdependence, sharing, and trading of concepts, language, and metaphors between the natural sciences and the dominant social-economic order. The concepts of "struggle for existence" and "survival of the fittest," central to Darwinist theories of natural selection, were also essential for the laissez-faire philosophy of a new, rising, competitive capitalism. (1985:21)

In *Why Feminism?* (1999), psychologist Lynne Segal devotes an entire chapter to Darwinism and especially to evolutionary psychology ("Few things are more depressing for me to have to write about than the renaissance of Darwinian fundamentalism," she remarks on page 80). In appealing to postmodernism and rejecting any belief in universal laws of human nature, she attacks the "monetarist rhetoric" of the neo-Darwinian "tale" (1999:101). Hilary

Rose, on her part, describes Darwin's theorizing as "part and parcel of his times—the innovation lies in transferring a social theory into biological discourse" (2000:109).

The social theory Rose is referring to is that of competition over scarce resources within human populations, as developed by economist and demographer Thomas Malthus in his *Essay on the Principle of Population* (1798). Malthus's *Essay* made an end to the optimistic belief of his days that population growth automatically meant more prosperity. He argued that, since the number of people grows much faster than the amount of food, problems will inevitably arise. The population will always automatically be reduced because of starvation. Malthus's *Essay*, and above all his observation that in nature, too, plants and animals have more offspring than can possibly survive, was instrumental in the development of Darwin's evolutionism (Braeckman 2001).

Does this mean that Rose and the other critics are right in suggesting that Darwin was nothing but an ideologue? Do the premises really justify the conclusion? Biologist John Maynard Smith asks the same question:

> Both Darwin and Wallace[3] have told us of their debt to Malthus. The formal parallels between the theory of natural selection and the competition implicit in a capitalist society are obvious. Must Darwinism therefore be rejected by socialists? . . . The Marxist scientists of my youth—Bernal, Haldane, Levy—drew a distinction between science on the one hand and other human intellectual constructs, for example, theology and law, on the other. They saw science as genuine knowledge about the world, although they emphasized the distortions arising from society. (1997:523)

His answer is unequivocal: "To reject Darwin because some of his ideas came from an analogy with the capitalist society in which he lived, is, precisely, the kind of thing we should not allow our prejudices to lead us to" (1997:523).

As philosopher Johan Braeckman (2001) notes, the crucial point is whether Darwinian theory can provide a scientific explanation for the phenomena it purports to explain, irrespective of Darwin's sources of inspiration. And it can. Braeckman cites historian Peter Bowler,[4] who argues that, indeed, natural selection was not simply an induction made on the basis of empirical facts, but neither was it just a reflection of the competitive ethos of Victorian capitalism. Clearly both scientific and nonscientific factors played a role in shaping Darwin's thoughts. He let himself be inspired by many sources and synthesized them into a unique explanatory model of the origin of species. According to Bowler, he used his scientific and cultural sources of inspiration in such a highly creative way that the resulting theory by far outstripped his contemporaries in imaginative skill. Philosopher Daniel Dennett (1995), too, acknowledges that Darwin, who was after all just a human being like the rest of us, was influenced by the values and concepts of his time. But, as Dennett points out, the metaphors of economy, that come to mind so readily when one is talking about evolution, *derive their power from* one of the basic characteristics of Darwin's discovery.

The fundamental processes underlying evolution by selection have been observed many times in the laboratory and in the field and have never been countered by a single study or finding, so that evolution by selection is viewed by many biologists as a fact. The theory has those characteristics that scientists seek in a scientific theory: It organizes known facts about organic life, it leads to new predictions, and it provides guidance to important domains of scientific inquiry (Buss 1999).

Thus we must conclude that far from being a "social theory being transferred into biological discourse," evolution by natural selection is a firmly established scientific theory, which cannot in the least be undermined by linking it to capitalism or to other value systems, whether by feminists or by science critics in general. It can even be said that evolution by natural selection is now established as a *fact.* This does not mean, of course, that everyone agrees about the details of the process. The fact that Darwinian theory has undergone some important modifications in the course of the twentieth century and that fierce debates are going on about, for instance, the exact role of natural selection relative to other agents of evolutionary change (e.g., Dennett 1995; Gould 1997; Gould and Pinker 1997; Wright 1990, 2000) only testifies to the self-correcting and self-critical nature of the scientific enterprise. A discussion of the more technical details of evolutionary theory does, however, not belong to the scope of this work.

Social Darwinism

A comparable and sometimes simultaneously used feminist strategy of discrediting Darwinian theory is equating it with social Darwinism, here interpreted as the view that social inequality between the sexes, classes, or races is the product of natural and sexual selection and is therefore right. Taking measures to reduce that inequality means impeding the progress of the species and hence should be avoided. Women's function, in this line of thought, is primarily that of childbearing.

The simple and most common answer to this claim about Darwinian theory amounting to social Darwinism is that it concerns two very disparate concepts. The first is a scientific theory; it tries to explain how the natural world works. The second is an ideological doctrine; it justifies colonialism and laissez-faire conservatism, and does this, moreover, in a way that has less to do with Darwin than some people seem to think. The leading theorist of social Darwinism, the sociologist and philosopher Herbert Spencer (1820–1903), combined Lamarckian ideas with Darwinian ones, proclaiming that progress is inherent to evolution, something that Darwin did not agree with. Spencer believed that competition between organisms causes adaptations, which are subsequently inherited by the organism's offspring (Braeckman 2001). In general, a better term than 'social Darwinism' might be 'social Spencerism' or 'social Lamarckism'. Darwin considered Spencer's work much too speculative and obscure and, apart from borrowing his phrase 'survival of the fittest', distanced himself from Spencer's writings (Browne 2002).

The more complex account of the relationship between Darwinian theory and social Darwinism is that for Victorian scientists the relevance of biology for social theory and policy went without speaking; indeed, the term 'social Darwinism' would have baffled Darwin (Paul 2002). Social Darwinism, in this sense, is just the then self-evident search for the social and political implications of evolutionary theory. And in the late nineteenth century this theory was to many people a mixture of the work of Darwin, who stressed the primacy of the mechanisms of natural and sexual selection, neo-Lamarckian theories, explaining evolution by the inheritance of acquired characteristics, and the work of Herbert Spencer. Those who wanted to apply evolution to social behavior thus could pick and choose from these and other theories the aspects that appealed to them ideologically and then construct their own theoretical framework (Hermans 2003). Furthermore there was Darwin's own ambivalence in his writings, as well as the fact that, strictly speaking, Darwinian theory can be used to accommodate almost *any* political line of reasoning. Indeed, in the late nineteenth century it appealed to socialists as much as to defenders of laissez-faire, which means that socialists were social Darwinists, too, in the broad sense of the word. It just depends on what elements you prefer to focus on in Darwinian theory: cooperation or competition.

Darwin dissociated himself from social Darwinism in its strict sense—the sense in which it tends to be interpreted by biologists and philosophers of science—something that Rose (2000) does acknowledge. Bleier, however, continues the quote on Darwinism and capitalism as follows:

> The other side of the coin—the implication of the disastrous effects of the weak, the inferior, or "degenerate" on the survival and vigor of the species or, more usually, the "race"—became scientific bases for proposals for eugenics programs in the United States and England in the nineteenth and twentieth centuries and for the ultimate program of extermination of inferior (i.e., non-Aryan) peoples in Europe in the 1930s and 1940s. (1985:21)

Bleier thus makes a straight inference from Darwinism to fascism, without providing any justification for this kind of "logic." A sober reading of Darwin's work reveals that it takes a huge leap of the imagination to pass the responsibility for the evils of fascism onto Darwin. As to eugenics, this program was not introduced by Darwin, but by his nephew, Francis Galton, and never defended by Darwin. Bleier, however, does not make this clear.

This kind of misleading argument can be countered in the same way as the 'Malthus-argument': A scientific theory has to be evaluated on its scientific validity. Nobody will be better off by equating Darwinism with a pernicious ideology, neither feminists, who only lose their intellectual credibility in informed people's eyes, nor their audience, which, misinformed, runs the risk of turning its back to science.

Evolution by Sexual Selection

Before discussing Darwin's theory of sexual selection, it might be useful to explain what is meant by sex itself. Sex is the process of combining genes from more than one source. It exists among viruses, bacteria, and all major groups of higher organisms. Among most higher organisms, we find two distinct types—male and female—contributing genetic material. In mammals, the gamete (sex cell—egg or sperm) of each sex contains a single or haploid set of chromosomes. The fusion of these gametes creates a new, genetically unique diploid organism. An individual offspring thus inherits genes from two parents, but is identical to neither. Important to know is that biological sex is determined simply by the size of the gametes: The organisms with the larger gametes, or ova, are females, and those with the smaller gametes, or sperm, are males (Birkhead 2000).

The question of why sexual reproduction has ever evolved at all has been one of the most persistent puzzles within evolutionary biology. Many species, after all, such as certain species of mites, reproduce asexually, with the offspring being identical copies of the parents (with the exception of whatever mutations occur). Sexual reproduction could never have evolved if there had not been some reproductive advantage to it, relative to asexual reproduction; such is the logic of evolution by selection. The leading explanatory theory, developed by evolutionary biologist William Hamilton and biologist Marlene Zuk (1982), holds that parasites are responsible for the origins of sex: Since sexual reproduction provides genetically diverse offspring, parasites cannot spread as easily from host to host as in the case of genetically identical offspring. Thus more offspring can survive.

In *The Descent of Man and Selection in Relation to Sex* (1871/1874) Darwin defined sexual selection as selection caused by the reproductive competition between members of the same sex and species. As he observed, in most species with sexual reproduction, members of one sex, usually the males, compete between themselves for mating access to the other sex. They do this by way of threats, combats, and useful weapons such as antlers, sharp claws, and strong muscles, but also in a far less intimidating way: through aesthetic display. They show off with beautiful colors, with conspicuous behavior, or with elaborate songs. Darwin posited sexual selection as a way to account for many of these conspicuous physical and behavioral traits in males that could endanger their survival.

The evolution of the peacock's elaborate tail, or of male animals displaying for female ones, for example, is difficult to explain by the process of natural selection as Darwin described it in *On the Origin of Species.* These traits are so energy demanding and so likely to make the animal vulnerable to predators that natural selection would have normally selected them away in an early evolutionary stage. According to Darwin, the reason this did not happen is female choosiness: Male ornamentation and competition for females evolved because females prefer to mate with the strongest and best-ornamented males. Female choice thus influences the course of evolution: The chosen males will have a greater reproductive success relative to the unsuccessful ones. Over evolutionary time, their traits will spread through the population and become evermore exaggerated, since females will continue to choose the strongest and best-ornamented males (Darwin 1998 [1871/1874]).

The "prize" of the winners in the sexual contest is thus not survival, but leaving more offspring. In extreme cases of sexual selection, such as with the elephant seals, the number of losers is huge and that of winners very small: 5 percent of the males sire 85 percent of all offspring produced in a breeding season (Buss 1999).

Darwin made a correct observation about female choosiness in most sexually reproducing species. He could not, however, answer the question of why it is the females that are generally the choosiest or why there is choosiness at all. It would take the work of evolutionary biologist Robert Trivers (1972) to solve this puzzle. We see the exact explanation for female choosiness later in this chapter. Here I concentrate on the trouble with Darwin's view of females.

Darwin and the "Coy" Female

As I argued before, Darwin's claim that female choice had shaped striking male ornaments and behavior over evolutionary time was in fact quite revolutionary—apart from the already revolutionary character of Darwinian theory as a whole. To be sure, male competition was readily accepted by his evolutionarily minded contemporaries, but female choice as a driving force in the evolutionary process was not. As Darwin himself said:

> Most . . . naturalists . . . admit that the weapons of male animals are the result of sexual selection—that is, of the best-armed males obtaining most females and transmitting their masculine superiority to their male offspring. But many naturalists doubt, or deny, that female animals ever exert any choice, so as to select certain males in preference to others. (Darwin 1882,[5] as cited in Cronin 1991:114)

Darwin applied sexual selection theory very liberally; he detected female choice "at the drop of a feather" (Cronin 1991:243). Reaction against the theory went so far, however, that after Darwin's death it was almost completely neglected for nearly a century. The few people who did recognize its value, such as geneticist and mathematician Ronald Fisher, were fiercely criticized (Miller 2000b). The reasons for the neglect are diverse, but an important aspect of it was the uneasiness over female choice of mates. It was assumed that females are passive in the mating process. Conspicuous male traits, it was assumed, evolved just for the sake of species recognition.

Only in the 1970s would scientists gradually come to accept the profound importance of female choice in the animal world, and only in the 1980s would scientists begin to document within our own species the active strategies that women pursue in choosing and competing for mates (Buss 1994). Meanwhile the agency that Darwin had conferred to females, however slight it may have been when compared to current theorizing on sexual selection, was largely forgotten.

What exactly did Darwin write about female choice and about the nature of the sexes? Some representative quotes make clear what Hrdy means when she notes: "Guided by a theory that was not only powerful but largely correct, Darwin had an uncanny knack for separating the anecdotal chaff from the true kernels of natural history . . . but when it came to females, and especially female choice, his vision was impaired by the blinkers of Victorian prejudice" (1997:8):

> When the organs of sense and locomotion are . . . more highly developed in the one [sex] than in the other, it is, as far as I can discover, almost invariably the male which retains such organs, or has them most developed; and this shows that the male is the more active member in the courtship of the sexes. The female, on the other hand, with the rarest exceptions, is less eager than the male. . . . She is coy, and may often be seen endeavoring for a long time to escape from the male. . . . the female, though comparatively passive, generally exerts some choice and accepts one male in preference to others. . . . The exertion of some choice on the part of the female seems a law almost as general as the eagerness of the male. (Darwin 1998 [1871/1874]:229–230)

> Man is more courageous, pugnacious and energetic than woman, and has a more inventive genius. (Darwin 1998 [1871/1874]:576–577)

> Woman seems to differ from man in mental disposition, chiefly in her greater tenderness and less selfishness. Man is the rival of other men; he delights in competition. . . . It is generally admitted that with woman the powers of intuition, of rapid perception, and perhaps of imitation, are more strongly marked than in man; but some, at least, of these faculties are characteristic of the lower races, and therefore of a past and lower state of civilization. The chief distinction in the intellectual powers of the two sexes is shown by man's attaining to a higher eminence, in whatever he takes up, than can woman—whether requiring deep thought, reason, or imagination, or merely the use of the senses and hands. (Darwin 1998 [1871/1874]:583–584)

> Thus man has ultimately become superior to woman. (Darwin 1998 [1871/1874]:585)

Before and after Darwin, more enlightened views of woman have been written down, to put it mildly. Clearly he did not succeed here in separating his scientific attitude from his social prejudices. As a scientist, the least he should have done was to consider the ways in which women's freedom and capabilities were being socially curtailed, instead of taking the plight of women for granted. But then again, in doing so Darwin only complied with the traditional wisdom of his time and class. It would take the feminist movement to undermine some age-old prejudices about women.

As to female choice within the (nonhuman as well as human) animal kingdom, it has now become clear that Darwin's conception of it was much too narrow, as was his conception of sexual selection. First, he thought that in modern society it was males rather than females who exercise sexual choice, and he regarded this as a sign of progress. Second, in current theorizing, female choice has become much more active, and female choice and male-male behavioral contests are seen as just two of the mechanisms of sexual selection. Male choice of mates, female-female competition, forced copulation, and aggressive conditioning of female behavior are among the other ones, as well as postcopulatory strategies: sperm competition and sperm choice.[6] The current view of the sexes, and the important role women researchers have played in establishing it, will be further elaborated in chapter 5.

The exact extent to which in Darwin's view the female exerts a choice is not completely clear: She "generally exerts some choice," but she is also "comparatively passive" (1998 [1871/1874]:230). What is one to make of this? Darwinian feminists differ in their interpretation. Hrdy (1997) explains the contradiction (females making a choice and being passive at the same time) by saying that Darwin just concurred with Victorian wisdom about women, which held that the sexual feelings of the majority of them did not amount to much. According to Laurette Liesen (1995a), Darwin made it clear that females play only a minor role in the explanation of male traits, since female choice occurs only after males have competed with each other. Patricia Gowaty, however, thinks that to Darwin female choice was as important as male-male competition. Pondering the neglect of female preferences after Darwin's death, she adds: "I consider the long-standing theoretical primacy of male-male competition to be one of the most potentially misleading notions in evolutionary biology" (1992:229).

Gowaty thus seems to concur with what I suggested earlier: that Darwin attributed a far more important evolutionary role to females than evolutionary biologists would do for almost a century after him. Female choice would be rejected, but the appellation "coy" was to remain unchallenged dogma for the succeeding hundred years. Even today the term has not disappeared from evolutionary writings. We can find it, for instance, in Robert Wright's *The Moral Animal* (1994b), an introduction to evolutionary psychology.

What is inappropriate about the term "coy"? For one thing, it is plainly wrong. During the past three decades, research has revealed that the females of most species are anything but passive or sexually coy. Through their mate choices, they direct the course of evolutionary change—at least, to the extent that their choice is not thwarted by males (Birkhead 2000; Fedigan 1997; Gowaty 1997b, 2003; Hrdy 1997, 1999a, 1999b; Mesnick 1997; Smuts 1995, 1996). A primary causal factor in this greater attention to female interests and strategies was the rise of women in the field of animal behavior. Partially because of them, new questions were being asked and new answers given.

Yet the females of most sexually reproducing species *are* more discriminating than males about mating. There is a huge difference, however, between "coy" and "discriminating": The former is not a value-neutral term. It is laden with sex-linked cultural meanings and as such does not belong within a scientific vocabulary. As Helena Cronin notes:

> I can't resist wondering what words would be used if the sex-roles were reversed. Would a (male) investor or business executive be called coy for not rushing headlong into the first option? If males were choosy about mates, would they be "coy"—or discriminating, judicious, responsible, prudent, discerning? (And, by the way, would females be "eager"—or would they be wanton, frivolous, wayward, brazen?) (1991:248)

Antoinette Brown Blackwell: The Road Not Taken

Antoinette Brown Blackwell was the first woman to publish a critique of *The Descent of Man*, four years after its publication. Darwin (and Spencer), she argued in *The Sexes throughout Nature* (1875), had not given enough attention to the role of females in natural and sexual selection. As she writes: "With great wealth of detail, he [Darwin] has illustrated his theory of how the male has probably acquired additional masculine characters; but he seems never to have thought of looking to see whether or not the females had developed equivalent feminine characters" (1875:16). She knows that, as a woman, she will be considered presumptuous for criticizing evolutionary theory, but she sees no alternative: "Only a woman can approach the subject from a feminine standpoint; and there are none but beginners among us in this class of investigations. However great the disadvantages under which we are placed, these will never be lessened by waiting" (1875:22).

Blackwell thus identifies Darwin's limited perspective as a male observer as one of the main problems of the theory of sexual selection. For all male secondary sexual characteristics that Darwin describes, there are equally important corresponding traits in females, she stresses. The more complex an organism, the greater the division of labor between the sexes. The net effect of the complementations, however, leads to sexual equality.

Her view of the sexes was, as was Darwin's, influenced by Victorian values, but it was less superficial than his when it came to women. As Blackwell sees it, the greater sexual passion, aggression, and constructive intellect of the male is balanced, among others, by the deeper parental and conjugal affection of the female and by her higher intellectual insight and greater facility in coping with details. She does not doubt the correctness of her inference that sexual equality is inherent in evolution. The precise character of these complementary male and female characteristics is, however, advanced more as a hypothesis. As she says, "[t]he facts need careful investigation" (1875:128).

Blackwell thus writes in a scientific spirit, thinking logically, weighing the evidence. Although part of her ideas turn out to be ill-founded, such as her Lamarckian-Spencerian belief in the progress of evolution, her reasoning was methodologically sound. It is a great pity, therefore, that her critique was not heard and that, as Hrdy writes, "[h]er contribution to evolutionary biology can be summed up with one phrase: the road not taken" (1999a:13). "This turning point," Hrdy says, "over a century ago, left a rift between feminism and evolutionary biology still not mended." The focus in evolutionary theorizing would for long remain mainly on the supposed pivotal role of male-male competition in the evolution of a species, with females coyly safeguarding their fertility until the appropriate male partner arrives and after that being preoccupied with motherhood. Hrdy continues:

> Small wonder, then, that audiences sensitized to both the excesses of social Darwinism and conventional sexism have found this emphasis upsetting. Yet evolutionary biology, and its offspring, sociobiology, are not inherently sexist. The

> proportion of "sexists" among their proponents is probably no greater than the proportion among scientists generally. (1999a:14)

In evolution, Blackwell finds a basis for demanding more freedom for women:

> Evolution has given and is still giving to woman an increasing complexity of development which cannot find a legitimate field for the exercise of all its powers within the household. There is a broader, not a higher, life outside, which she is compelled to enter, taking some share also in its responsibilities. (1875:135)

As to the question of the relationship between evolutionary knowledge and politics, it must be clear that *no* direct inference can be made from facts to values, from nature to ethics and politics. It must also be clear that, in order to make efficient political decisions, knowledge of human nature, that is, of human needs and propensities, is indispensable. Indeed, a political ideology is *always* implicitly or explicitly informed by a view of human nature, even if that view holds that it is the nature of humans to have no nature. There is thus an *indirect* link between human nature and politics. As Blackwell was probably the first to show, evolutionary knowledge can be used in a liberating way, to argue for social equality between the sexes.

Reading feminist accounts of Blackwell's critique, however, one is struck—again—by the authors' inability to distinguish between the scientific and the ideological character of a theory. Sue Rosser mentions Blackwell, Wallace, and Darwin himself as persons raising doubts about some key points of the theory of sexual selection. "Why, then, one wonders, did Darwin insist on the theory so much?" she asks, concluding that it must have been for ideological reasons:

> In order to make the differentiation between males and females as strong as possible, the theory of sexual selection is needed. The theory is the agent of differentiation, that which assures an ever-increasing separation between the sexes and their operation in two quite distinct realms that only touch for the purpose of procreation. (Rosser 1992:58)

Would Rosser from the resistance against female choice also infer that Darwin insisted on it only for ideological reasons? It is a strange kind of logic to conclude that, because a scientist has troubles with some aspects of his theory, he should reject it, and that if he does not do so, this means that the *whole* theory is an ideological construct. Moreover, since Rosser seems to agree that some of the doubts raised about sexual selection may have hit the point, she seems to imply that there might be a theory that is scientifically *right*. Something can only be wrong measured against things that are right. But how can she know that if she rejects the existing standards of scientificity, as we saw she does in the previous chapter?

Anne Fausto-Sterling's (1997) account is a comparable one. She describes with admiration Blackwell's chiding of Darwin and especially her conclusion that evolution dictates that men should prepare the food. Subsequently she cites Eliza Gamble, another Victorian female writer who criticized Darwin. In her book *The Evolution of Woman: An Inquiry into the Dogma of Her Inferiority to Man* (1894), Gamble argued that evolution proved female superiority. Fausto-Sterling detects traces of Victorian values in Gamble's work, but for the rest she does not seem to have any trouble with the kind of inferences Blackwell and Gamble make on the basis of evolutionary theory. Although she definitely condemns any sociobiological approach to humans ("the entire process strikes me as misguided at best, socially pernicious at worst," Fausto-Sterling 1997:48), not one word of criticism is uttered when women use the theory to their own purposes. She calls it "tactics" (Fausto-Sterling 1997:46).

One wonders how this kind of inconsistency within feminist theorizing can ever be constructive. If a theory is bad science, why not make it better science, instead of rejecting the whole of it as an ideological construct, but still making use of it when it suits one down to the ground?

The 1930s and the Rise of Ethology

With the 'Modern Synthesis' of the 1930s and 1940s, that is, the fusion of Darwin's theory of evolution by natural selection with the discovery of genetic inheritance, Darwinian theory was put on a firmer footing. In the 1930s, the first major discipline around the study of behavior from an evolutionary perspective was formed: the field of ethology, the study of animal behavior from an evolutionary perspective. Darwin, after all, envisioned his theory of natural selection as being just as applicable to behavior as to physical structures. That behavior is not exempt from the sculpting hand of evolution is suggested by the observation that all behavior requires underlying physical structures and that species can be bred for certain behavioral characteristics using the principle of selection (Buss 1999).

One of the first phenomena documented in this new branch of evolutionary biology was imprinting. Ducklings form an association with the first moving object they observe in life—usually the mother—and follow it wherever it goes. Imprinting is clearly a preprogrammed form of learning and part of the evolved structures of the duckling's biology. Ethologists studied the proximate or direct *and* the ultimate or evolutionary mechanisms of behavior. Proximate mechanisms were, for instance, the movement of the mother and the events during the duckling's lifetime that cause changes. The ultimate function or adaptive value of behavior was, for instance, to keep the baby duck close to the mother, which helps it to survive (Buss 1999).

Ethologists, however, often did not go very far in explaining behavior. They still tended to describe observable behavior rather than to look for the underlying mechanisms responsible for generating it, and they did not develop rigorous criteria for discovering adaptations (Buss 1999). As we saw in chapter 2, at that time the study of animal behavior was also still very male biased: The focus was on male dominance hierarchies, which were supposed to be the basis of social organization. Females were not considered very interesting. Now that we know more about the reception of Darwin's theory of sexual selection, it should be easier to understand why females escaped the attention of (male) ethologists: After Darwin's death, female agency had been removed from "the mask of theory." Neither should it surprise us that it was mainly through the work of women *primatologists* that this bias was finally corrected in the 1970s. As indicated in chapter 2, primates seem to be the only nonhuman animals that generate so much empathy on the part of the investigator that he or she starts to focus on his or her own sex. With the rise of women in primatology, some major theoretical changes were thus more to be expected than, say, with the rise of women in entomology.

Women in Primatology

Because of the important role of women in primatology, the field is often singled out for praise by feminists, and sometimes interpreted as a feminist science (Fox Keller 1982; Rosser 1992; Segal 1999). This observation elicits the following remark from primatologist Linda Fedigan: "What strikes me as curious . . . is the possibility that theirs is an unrequited love affair, because there are so few primatologists who acknowledge that they are feminists or admire, much less pursue, feminist goals" (1997:58).

Is primatology a feminist science? In an article by the same name, Fedigan (1997) ponders this question. As we saw in chapter 2, however, feminist science, if it is really science, is just traditional science done by feminists, and thus the appellation 'feminist' is superfluous.

A better question would be: How much did feminism have to do with this transformation in primatology (and, as a consequence, in sociobiology)? The answer is: a lot, but mainly in an *indirect* way. The primary impact of feminism was that it led to more women studying and contributing to biology. Feminism also led to a world in which men were faced with women who took control of their own lives. As a consequence, 'female choice' did not seem so strange anymore (Miller 2000b). The result of more women entering the field was a widening of the theoretical framework to include female interests and strategies. With the entrance of women researchers this change was to be expected, since in the field of primatology gender is intrinsic to the subject being studied. Feminism as such seems to play only a minor role on this level.

Darwinian feminist Hrdy sees it as follows:

> Feminism was part of the story, but not because women primatologists and biologists had different sensibilities than male scientists, or because feminists do science differently. Rather, women fieldworkers were predisposed to pay more attention when females behaved in "unexpected" ways. When, say, a female lemur or bonobo dominated a male, or a female langur left her group to solicit strange males, a woman fieldworker might be more likely to follow, watch, and wonder than to dismiss such behavior as a fluke. Women were also more likely to have been affected by feminist ideas. My own interest in maternal strategies grew directly out of an emphatic identification with my study subjects. . . . Women scientists were less likely than male scientists to identify with authority and with the scientific status quo. Women fieldworkers . . . may have been more willing to entertain unorthodox ideas about sex roles. . . . But feminism per se had little to do with the conclusions I reached. (1999a:xviii–xix)

She concurs with the view that feminist science does not exist when she says:

> A few have gone so far as to ask whether the incorporation of "the female perspective" into evolutionary biology should be read as a "triumph" for science or for feminists. I think such questions are mischievous. The answer is so obvious: any time wrong ideas are corrected, science wins. If biases were there in the first place because of sexism, and a feminist perspective helped us to identify them, it is still science that comes out ahead when they are corrected. (Hrdy 1999a:xix)

What kind of conclusion does Fedigan reach? She is in doubt. She points out that the nearly equal proportions of women and men primatologists are not particularly striking from the perspective of the behavioral sciences, such as anthropology and psychology. Yet it does seem, she continues, that women have been more responsible than men for developing our present model of the female primate, which stresses female sexual assertiveness, female bonding, female competition, and female social strategies. Primatology has shown itself to be very responsive to criticisms of androcentric language and interpretations, and quite willing to redress the past focus on male behavior with a present focus on both sexes and on the relationship between the sexes. "If this is a feminist transformation, then it has happened in primatology" (Fedigan 1997:67).

Fedigan thinks we should give credit to the feminist critique for drawing our attention to androcentric bias in science. She is, however, not sure that it is really due to feminist ideology that we have come to a better understanding of female primates. The patient observation needed to study these complex, long-lived animals in the wild, for instance—a practice that was uncommon in the early years of ethology—leads to a very different attitude than that developed when studying short-lived creatures in the laboratory. Fieldwork has made clear that many primate societies are female bonded, something that may facilitate a strong focus on females. She concludes:

> [S]ome of the trends in primatology that have been identified in this chapter, such as the development of a female as well as male perspective, and the move from reductionism and dualisms to increasingly complex, sophisticated explanatory models, could arguably result from the processes intrinsic to the maturation of all scientific disciplines. It is possible that any new science would go through initial stages of being relatively mechanistic and making simplifying assumptions. As the science matures, it would be expected to graduate to sophisticated models that are more complex, dynamic, and multi-factorial. Thus, the goals of those who try to develop a better, more mature science of primatology may sometimes dovetail with the goals of those aspiring to a feminist science. (Fedigan 1997:70)

The 1960s and Inclusive Fitness Theory

Inclusive fitness theory or kin selection theory was formulated by evolutionary biologist William Hamilton in the early 1960s. The theory transformed the entire field of biology. In essence it meant a change from an organism-centered view to a gene's-eye view. Classical Darwinian explanations, before the 1960s, focused on survival and reproduction from the point of view of the individual organism or the group. That was the level at which natural selection works, it was thought. Modern Darwinian theory is about "selfish" genes, as Richard Dawkins (1976) famously called them. In practice, however, natural selection at the level of genes often means the same as "at the level of the organism," since a successful strategy of a selfish gene is very likely to promote the survival and reproduction of the organism that houses the gene (Cronin 1991).

Hamilton reasoned that natural selection favors characteristics that cause an organism's genes to be passed on, *regardless of whether its offspring is produced directly*. An organism's genes are also housed by its genetic relatives: It is genetically related by 50 percent to its brothers and sisters, by 25 percent to its grandparents and grandchildren, by 12.5 percent to its first cousins, and so on. Not only is its fitness thus enhanced by producing offspring directly; it can also increase the reproduction of its genes by helping relatives to survive and reproduce. Hamilton called the sum of both kinds of reproductive success 'inclusive fitness'. Adaptations, which may be defined as evolved solutions to a specific problem of survival or reproduction, arise by this process. Genes that have effects that increase their reproductive success will replace other genes, producing evolution over time (Buss 1999).

Genes, of course, are not really selfish; they have no consciousness or intentionality. It is just a metaphor Dawkins thought fitting. The gene's-eye view now predominates in evolutionary biology and has been a stunning success. It is finding answers to questions such as how life arose, why there are cells, why there are bodies, why there is sex, why animals interact socially, and why there is communication. It has become indispensable to researchers in animal behavior (Pinker 1997).

The 1970s and Parental Investment Theory

Evolutionary biologist Robert Trivers is famous for a number of seminal papers, but the one most of interest to us here is "Parental Investment and Sexual Selection" (1972). By clarifying the relationship between Darwin's theory of sexual selection and parental care, Trivers succeeded in solving the problem of why females typically had evolved to be more choosy than males. The article brought female choice back to the foreground.

Part of the answer is anisogamy, from the Greek *aniso* (meaning unequal) plus *gametes* (eggs and sperm): gametes differing in size. As said earlier, the female is, by definition, the sex that produces the larger gamete. In the late 1940s, laboratory experiments revealed the relevance of this gamete dimorphism to sexual selection. Biologist Angus Bateman showed that a male fruit fly's fitness (defined here as the number of healthy offspring) rose in an approximately linear way along with the number of females he mated with, whereas a female gained little by copulating more than once. This asymmetry, he concluded, is caused by the limited egg production capacity of the female (Trivers 1972).

Trivers expanded Bateman's work. One consequence of eggs being larger and much more nutritious than sperm is that in cases in which internal fertilization has evolved—as it has done many times independently—it almost invariably occurs within the female. This in turn sets the stage for the evolution of mammalian pregnancy and lactation. Thus the reproduction of a female mammal is not limited merely by her egg production capacity, but also by the time and energy devoted to pregnancy and lactation. In taxa such as mammals, females make, as Trivers called it, a greater parental investment in each individual offspring than males, defining parental investment as "any investment by the parent in an individual offspring that increases the offspring's chance of surviving (and hence reproductive success) at the cost of the parent's ability to invest in other offspring" (Trivers 1972:139). The minimal cost of successful reproduction for a female mammal is thus a large investment of time and energy in pregnancy and lactation. The male, however, often invests nothing more than some sperm. He can garner the same fitness for no more parental investment than the cost of a copulation.

Before continuing, it is important to note that no conscious intent is implied on the part of the animals described, as it never is within evolutionary discourse. The point is that behavioral strategies that increase an organism's fitness relative to others will spread through the population and become evolutionarily successful.

Various sex differences in sexual behavior follow from the asymmetry in initial parental investment, Trivers showed. For males more than females, reproductive success is limited by the number of fertile partners they mate with. For females more than males, on the other hand, reproductive success is limited by the time and energy required to successfully raise offspring. Males therefore are usually more eager than females to mate at any time with any partner who may be fertile, whereas females are usually more careful than males to choose mates who seem likely to provide good genes, protection, parental care, or resources. This basic asymmetry sets the stage for the evolution of male exploitation of the female's reproductive efforts, since the male can disappear after conception in order to fertilize other females (Trivers 1972). Again, no conscious strategizing is implied.

Since females are relatively more choosy, males will compete with same-sex rivals for mating opportunities. Resulting male adaptations include not only weaponry such as antlers, but also other physical and behavioral traits that advertise the male's genetic quality. Indeed, conspicuous secondary sexual characteristics such as an elaborate tail or bright colors have been found to be markers of genetic quality in animals, just as symmetry has been found to be such a marker in human and nonhuman animals. The reason is that factors such as harmful genetic mutations, parasites, and a lack of food inhibit the normal development of an organism. By being symmetrically built and by displaying conspicuous characteristics, the male animal demonstrates to the female that he is healthy, so healthy that he can afford these kinds of costly characteristics (Etcoff 1999). Female taste can thus be explained adaptively, which means that, ultimately, sexual selection falls within the scope of natural selection. It is because of natural selection that sexual selection acts the way it does.

Although Trivers reintroduced female choice in evolutionary theory, he did not entirely escape the androcentric views of his time. The 'mixed reproductive strategy' that he proposed for socially monogamous species (forming a pair bond but having extra-pair copulations as well) was supposed to apply only to males, because females were not thought to benefit much from being promiscuous. Only in the last few decades have the benefits of female promiscuity been established (see chapter 6).

Thus the pieces of the puzzle of natural and sexual selection seem to come together. One final question remains. What is the evidence that the principle of anisogamy—the suggestion that the production of eggs is energetically costly and that of sperm is cheap—is right? This is an important point, since in evolutionary biology anisogamy is the basis for explaining all other differences between the sexes.

Costly Eggs and Cheap Sperm: But Is It True?

In feminist literature we frequently find the principle of anisogamy questioned (Fausto-Sterling 1992, 1995; Hubbard 1990; Snowdon 1997; Tang-Martinez 1997). There is, of course, nothing wrong with questioning the interpretation of earlier scientific research; that is the way science makes progress. So let us have a look at what the critics say.

To recapitulate: The principle of anisogamy holds that eggs are very costly to produce, since they are large in size and contain large amounts of expensive nutrients, such as yolk. Sperm, on the other hand, are considered to be cheap, since they are tiny, compared to eggs, and generally are nothing more than a packet of chromosomes without nutrients. This concept of unequal investment, called 'Bateman's principle', was expanded by Trivers in 1972 to include a parent's entire investment in its offspring—including gestation, feeding, and protection. As a result, Trivers argued, females have evolved to be discriminating in selecting mates, whereas males have evolved to be relatively indiscriminating. This process in turn governs the operation of sexual selection (Trivers 1972).

"This premise," Zuleyma Tang-Martinez writes in reference to Bateman's principle, "has achieved dogmatic proportions in biology" (1997:129). She thinks it is wrong, arguing, as do Fausto-Sterling (1992), Charles Snowdon (1997), and Ruth Hubbard (1990), that one should compare the cost of one egg to the cost not of one sperm, but of one ejaculate, which contains *millions* of sperm. This sounds, indeed, like a very reasonable critique. Furthermore, as she points out, the costs incurred by males that must defend territories in order to mate and that engage in active combat prior to mating must be enormous. This, again, sounds plausible. She continues:

> The preponderance of the evidence at present suggests that a major tenet of modern evolutionary and sociobiological theory—namely, costly eggs and cheap sperm—has been based on what appear to be invalid and unwarranted assumptions. Scientists influenced by the sexual dynamics (coy females; sexually aggressive

> males) of modern, Euro-American societies apparently could not imagine that females in other species (and cultures) might behave differently. They, therefore, erected a theoretical framework (costly eggs and cheap sperm) to rationalize their biases. (Tang-Martinez 1997:130–131)

Now this interpretation seems rather rash. If it were true that evolutionary biologists only wanted to "rationalize their biases," then how can we explain that the principle of anisogamy still persists, despite the fact that the sexual assertiveness of females is now recognized as real by most evolutionary scientists (e.g., Zuk 2002; Birkhead 2000; Buss 1994, 1999; Cronin 1991; Gowaty 1997b, 2003; Hrdy 1999a, 1999b; Mealey 2000; Miller 2000b; Pinker 1997; Smuts 1995, 1996) and, certainly, despite the fact that some of these scientists are ardent feminists? Hrdy, for instance, writes about the critiques:

> Such objections are valid in that the theory of anisogamy has been used too deterministically and without adequate consideration of such important exceptions as those species of cockroaches and butterflies where male sperm is accompanied by various nutrients in addition to the genetic material, or the many birds and mammals where male care is essential to survival of offspring. Nevertheless, critics of the theory of anisogamy often suffer from an unrealistic assessment of just how difficult it is in the natural world to obtain the recources necessary to reproduce. . . . At their most extreme, I have heard such critics downplay as "trivial" the energetic costs of egg production, gestation, and even lactation. Such criticisms assume an infinite resource base and are not pertinent to mammalian evolution. (1999a:205–206)

Thus on Hrdy's account the critics were right in pointing out that the theory had to be applied in a more nuanced way, but this does not at all mean that the principle of anisogamy has become invalid. Hrdy wrote this originally, however, in 1981. Maybe the preponderance of the evidence really suggested otherwise in 1997? Fausto-Sterling definitely seems to think so when she writes that her critique has "turned out to be right on target," adding: "Few things have pleased me more than to see my and Hubbard's early prediction that there was something wrong with the anisogamy argument born out" (1995:172).

In saying this, Fausto-Sterling refers to Darwinian feminist Laurette Liesen (1995a), who acknowledges that studies of primates and of the hermaphrodite worm *Cuenorhabditis elegans* challenge the idea that male sperm is cheap and unlimited. Early descriptions failed to emphasize that this judgment is true only in comparison to female eggs, Liesen says. Indeed, behavioral ecologist Tim Birkhead (2000) concedes that for the majority of species we have absolutely no idea how much energy is needed for sperm formation. Studies suggest, he says, that the cost is not trivial, and that it may at least be equivalent to the costs of courting, copulating, and fighting rivals.

Does this observation, however, warrant Fausto-Sterling's conclusion? It seems not. As Hrdy stresses with regard to humans—but the argument seems to hold for mammals in general—"[w]hat is costly . . . is not the egg, or even the conceptus, but the *nurturing* it takes to rear a human being" (1999b:423, emphasis in original). The anisogamy argument is not only about costly eggs but also about the time and energy devoted to pregnancy and lactation. When all costs for both sexes are thus adduced, the asymmetry in investment seems to remain. The conclusion thus remains valid that females devote more of their total reproductive effort to parental effort, while males devote more of it to mating effort.

What are the indications that the theory of anisogamy is right on target, albeit that there might be many complexities to it that were not immediately recognized? First, it has been very successful in predicting the degree of sexual differences across animal species (Buss 1999). It obtains its strongest support, however, from the existence of a few species that exhibit 'sexual role reversal'. Although this may at first seem like counterevidence, it in fact emphatically *confirms* the theory.

Trivers's theory of parental investment and sexual selection, as based on the principle of anisogamy, made two profound predictions. First, individuals of the sex whose typical parental investment is greatest (usually, but not always, the females) will be more discriminating or selective about mating. Second, individuals of the sex investing less in offspring will compete among themselves for sexual access to the high investing sex (Trivers 1972).

There exist animal species where males intrinsically invest more than females, because the males are the ones delivering the nutrients or hatching the eggs. In these species we see the males being discriminating about mating, and the females competing with each other for access to males. This is the case with, among others, Mormon crickets, seahorses, pipefish, and Panamanian poison arrow frogs. These are species in which males provide more parental effort than females. The male Mormon cricket produces a large spermatophore (a capsulized mass of sperm) that is very nutritious. In areas where food is scarce, large spermatophores become extremely valuable to the female, but simultaneously become difficult for the male to produce, because they require extensive food consumption. Females compete with each other for access to the males holding the largest spermatophores. The females that are chosen by those males for depositing their spermatophore contain 60 percent more eggs than females who are rejected. In seahorses the female transfers her eggs into the male's pouch, after which she leaves to feed and make more eggs. The male releases sperm directly into the pouch and fertilizes the eggs. His pouch thus becomes the brood chamber for as many as 1,500 seafoals (Birkhead 2000; Buss 1999; Hrdy 1999b).

In a group of shore birds known as phalaropes (sea snipes) the males incubate the eggs and care for the young. Females compete for males and males choose between females. Sexual selection therefore operates more strongly on females. As a consequence they are larger and more aggressive than males and have a brighter plumage. The evolutionary rationale for this pattern is that females have a potentially higher reproduction rate than males. Female phalaropes can produce a clutch of eggs each week, whereas the incubation time for males is close to three weeks. It therefore benefits the female reproductively to leave in search of another male once her clutch is laid, and she typically does (Birkhead 2000; Geary 1998).

Species with sex-role reversal thus testify to the correctness of Trivers's principle that the sex that invests most in offspring will be the more discriminating one. There is no biological law that dictates that it must always be the females who invest more, but in the 4,000 mammalian species and the more than 200 primate species, it is the females, not the males, who undergo internal fertilization and gestation. In more than 95 percent of these species, females can effectively provide all of the parental care and in fact do so. High levels of biparental care are found when food resources are so scarcely distributed that the efforts of both parents are needed to raise the young successfully, as is the case in many bird species (Buss 1999; Geary 1998).

Several research groups are now putting the principle of anisogamy to rigorous tests, in species with sex-role reversal as well as in species with conventional male and female roles. The results so far indicate that the basic principle is sound, but that, as was also suggested earlier, there are intriguing complexities. In fruit fly seminal fluid, for instance, proteins have been found that increase the time that a female waits before allowing another male to copulate with her. This might indicate that female fruit flies are not holding back from mating out of choosiness, but because they are being manipulated by a male protein (Birkhead 2000; Knight 2002).

Other complications have arisen, such as the discovery of very large sperm (almost six centimeters!) in several species of fruit flies. Still the general conclusion seems to hold that the principle of anisogamy usually leads to male reproduction being primarily limited by access to females and to female reproduction being primarily intrinsically limited and affected by access to resources such as nutrients and nest sites (Birkhead 2000; Gowaty 1997b).

In this chapter I have laid out the basic principles of natural and sexual selection and dealt with the feminist critiques of them. Before proceeding to a consideration of the application of evolutionary theory to the human mind, it might be good to first devote some thought to the biophobia of many feminists.

Notes

Portions of chapter 3 have been previously published. Reprinted by permission of Sage Publications Ltd. from Griet Vandermassen, "A Tale of Male Bias and Feminist Denial," *European Journal of Women's Studies*, vol. 11, no. 1 (2004): 9–26 (© Sage Publications Ltd., 2004).

1. Angier 1999; Bleier 1985; Fausto-Sterling 1992, 1995, 1997, 2000a; French 1992; Gray 1997; Haraway 1991e; Hubbard 1990; Noske 1989; Sapiro 1985; Rose 1983, 1997; Rosser 1992; Segal 1999; Sork 1997; Sunday and Tobach 1985; Tang-Martinez 1997.

2. The idea of use-inheritance was, however, not originally Lamarckian; it had been around a long time before Lamarck picked it up. Neither was it the essence of his theory. Lamarckism was based on the thought that the workings of the laws of nature on living matter change according to the circumstances and that organisms have an inherent tendency toward more complexity. The main problem of Lamarckian evolutionary theory was its failure to provide a mechanism that can cause adaptations. Five decades later, Darwin proposed such a mechanism: natural selection (Braeckman 2001).

3. Alfred Russel Wallace (1823–1913) discovered the principle of natural selection independently from Darwin. At the time, Darwin's theorizing had already made much more progress than Wallace's. Both copresented the theory at a meeting of the Linnean Society in 1858.

4. Bowler 1990.

5. A Preliminary Notice: "On the Modification of a Race of Syrian Street-Dogs by Means of Sexual Selection." *Proceedings of the Zoological Society of London*, 367–369.

6. Birkhead 2000; Buss 1992, 1994, 2000; Cronin 1991; Ellis 1992; Gowaty 1992,1997b; Hrdy 1997, 1999a, 1999b; Mesnick 1997; Miller 2000b; Shields and Shields 1983; Smuts 1995, 1996; Thornhill and Palmer 2000; Thornhill and Thornhill 1983.

4

Biophobia within Feminism

> To set humans apart from even our closest animal relatives as the one species that is exempt from the influences of biology is to suggest that we do indeed possess a defining "essence" and that it is defined by our unique and miraculous freedom from biology. The result is an ideological outlook eerily similar to that of religious creationism.
>
> Ehrenreich and McIntosh 1997:12

CRITICIZING FEMINISM IS NOT A RISK-FREE ACTIVITY. Even dedicated feminists, such as Christina Hoff Sommers and Cathy Young, are accused of 'feminism bashing' whenever they dare to question some of the ruling tenets within academic feminism today. These tenets include a social constructionist view of knowledge, as we saw in the first two chapters, and an extremely environmentalist notion of the 'construction' of sex differences, as we see in this chapter. References to biology in the explanation of human traits tend to be dismissed as 'reductionist', 'biological determinist', and 'politically dangerous', not only in academic feminism, but also in the social sciences in general.[1]

There are some feminists who stress the human-animal continuity. Some do so because they defend animal rights and want to challenge anthropomorphism, such as the Dutch anthropologist Barbara Noske (1989). Others, such as the American science journalist Natalie Angier in her book *Woman: An Intimate Geography* (1999), believe that we can learn a lot about ourselves from studying our biological heritage, since we are just one species among others. Both Noske and Angier, however, fervently attack the use of evolutionary principles in the study of human behavior and the human mind, as is the case in sociobiology and evolutionary psychology. Noske seems to dismiss this approach out of hand—something that makes her argument about human-animal continuity inconsistent. Angier's line of reasoning is somewhat more consistent. She does not a priori condemn an evolutionary approach of human behavior; she only does so when she considers it male biased. The problem with Angier is that she too often relies on (androcentric) theories that had long been established as inadequate when she wrote the book.

Feminist biophobia has many facets and multiple causes. It will not work to dispose of the feminist critique of sociobiology and evolutionary psychology as merely being antiscience. Although a lot of it probably is, intelligent and stimulating books such as Angier's show us that it need not be so, and that the feminist suspicion of biological arguments about the sexes owes a lot to the tradition of androcentrism in science. This tradition has only been countered with the massive entrance of women in science. That does not mean that, today, male bias has disappeared from the scene. It does mean that whenever it arises, it will be critically analyzed and corrected by women researchers.

Sometimes, alas, it will be replaced by female bias. Christina Hoff Sommers (1994, 2000), Daphne Patai (1998), and Cathy Young (1999) have laid out the poor scientific level of some feminist research, due to its politicized character. Studies about sexual harassment, for instance, are often tailored to fit the preconceived radical feminist scheme of sexual harassment as a patriarchal tool for keeping women in subjection. That leads, among others, to the absence of serious research on the incidence of false accusations against men (Patai 1998).

Theories devised by women, whether they call themselves feminist or not, and whether they are working within the evolutionary sciences or not, need to be scrutinized as severely as those devised by men. It is my strongest belief that women can only benefit from setting themselves high standards of rationality and scientificity in deciding what research they consider to be reliable, instead of letting themselves be guided by ideological motives. Therefore I am convinced that the feminist biophobia, however understandable it may be, is counterproductive. Not only is it scientifically untenable, it is even unwelcome when consistently considered from a feminist point of view.

The former argument is also defended by feminist biologist Lynda Birke in her book *Feminism and the Biological Body* (1999), but for entirely different reasons: According to her, feminists need to focus on the body in order to analyze the ways in which physiological concepts are socially and culturally contingent. Birke argues for a nonmechanistic and a more descriptive approach, not only because she thinks mechanism gives a limited understanding of our bodies, but also "for more political reasons, because such a view of biology leads to determinism" (1999:136). "For most feminists, the idea that we are simply passively responding to controlling genes or organ systems, like puppets on strings, is anathema," she says (1999:164–165). But where are the biologists who assert such things? Birke does not do justice to the reigning interactionism in the biological sciences, probably because, to her mind, speaking of limits or constraints to what humans might do equals reductionism and determinism.

Fear of Sex Differences

In discussing feminist biophobia, I will use the term 'environmentalism' for the predominant view in feminism as well as in the social sciences in general that gender is "created and re-created, in response to specific tasks, roles and changing power relations" (Knapp 1998:368). Although I might also call this view 'social constructionism', confusion may arise with epistemological social constructionism. Therefore I prefer environmentalism.

It is not difficult to find expressions of this environmentalist position on gender; they abound within feminism. A few examples:

> Gender is social practice that constantly refers to bodies and what bodies do, it is not social practice reduced to the body. . . . Gender exists precisely to the extent that biology does *not* determine the social. It marks one of those points of transition where historical process supersedes biological evolution as the form of change. (Connell 1995:45, emphasis in original)

> [B]iology cannot be used to ground claims about "women" or "men" transculturally. (Nicholson 1994:89)

> [S]exual identity (and its normative ties with heterosexual desire) is only ever a precarious achievement. . . . Identities, of whatever sort, are . . . generated in social contexts which are nowadays always dynamic and shifting, involving a plurality of signifying practices and competing cultural narratives. (Segal 1999:158)

> Gendered characteristics—typically masculine or feminine behaviors, interests, or values—are not innate, not are they arbitrary. They are formed by historical circumstances. (Schiebinger 1999:72)

At the same time, the search for biologically based sex differences is made ideologically suspect. Rosser, for instance, writes: "One can imagine that a society free from inequality between the sexes would not view sex differences research as a valid scientific endeavor" (1992:71–72), by which she suggests that this research is motivated by the wish to sustain socioeconomic inequality between men and women. Similarly, biologist Ruth Hubbard (1990) believes that sex and gender dichotomies are set up as instruments of social control (so does Birke [1999]), and that our preoccupation with sexual matters is responsible for the fact that evolutionary theory is constructed around sex and procreation. Despite her argument to the contrary, however, the focus on the male-female difference seems to be a basic tendency in every human culture (Brown 1991). Humans automatically notice whether someone is male or female, something that is predictable from a Darwinian point of view, since partner choice is an evolutionarily vitally important matter.

Others argue that sex differences should be ignored because of the great deal of overlap between men and women. Anne Fausto-Sterling, for example, thinks that further research into sex differences in cognition is uncalled for: "If there are any differences at all, they are so small that intrasexual variation swamps them" (1992:221).

Biologically based explanations of sex-typical behavior seem to frighten many feminists to death. We see, however, that the environmentalist theories they put forward to account for behavioral differences between the sexes fail to explain these differences exhaustively. Before taking a look at these theories, we may ask why, indeed, we should study sex differences at all. Why not just ignore them?

According to neuropsychologist Doreen Kimura (1999), the answer is very simple: because they are there. Another important reason, Kimura argues, is that discovering the mechanisms for sex differences, such as those in the brain, often helps us understand differences between individuals of the same sex as well. The same applies to studying the relationships between sex-hormone levels and cognitive function. As to the objection that the overlap is more important than the differences, she replies:

> It is certainly true that, in the larger context of comparison with other species, the similarities between men and women far outweigh the differences. However, if we were to adopt the criterion of *no overlap* between groups as a requirement for accepting a difference, we would find almost no behavioral data in any field to be acceptable. (Kimura 1999:7, emphasis in original)

She offers the example of the effects of aging on memory. There will, of course, always be overlap in the scores of older and younger people. Yet we infer with some confidence that memory is, on average, not as good in the elderly. Even factors that have dramatic effects, such as damage to the brain, will often show a fair amount of overlap in the performance of those with and without brain damage. So, Kimura wonders, why should sex differences be treated differently from other kinds of data?

From the feminist point of view, the answer is self-evident: out of political motives (Fausto-Sterling 1992, 1997; Hubbard 1988, 1990; Nicholson 1994; Rosser 1992). This reasoning has a certain kind of logic to it. If, as proclaimed by many feminist critics, science always has a political agenda, then there is no reason why one should not try to model that agenda to one's own political ends. We have seen, however, that the existence of bad science or of politically induced scientific practices does not justify the conclusion that science as a collective enterprise is political. Therefore this kind of reasoning is untenable.

It is a very frustrating but ineluctable finding that some strands of academic feminism today have lapsed into dogmatism. I realize this is a heavy indictment. It is, however, the most adequate description of the feminist refusal to even consider the possibility that gender has something to do with biology. I am, of course, not the first feminist to point this out. Barbara Ehrenreich and Janet McIntosh, for instance, call "the trend—in anthropology, sociology, cultural studies and other departments across the nation—to dismiss the possibility that there are any biologically based commonalities that cut across cultural differences" a "new creationism" (1997:11). According to them, it is not simply a case of well-intended politics gone awry; it represents a grave misunderstanding of biology and science generally.

This trend is not limited to the United States; it extends to Europe as well (Brouns, Verloo, and Grünell 1995; Hermsen 1997; Michielsens et al. 1999; van Muijlwijk 1998). In what way exactly biology is misunderstood is explained later in this chapter. As to the reluctance to weigh the biological evidence on sex differences, I hand the floor over to Kimura:

> It is not only unjustified scientifically to take such a biased view of how behaviors are determined, it is contrary to common sense. The business of science is to find out how the world really works, not how it ought to work according to some wishful schema or other. Scientific explanations change as more information comes in, but at any one point in time a scientific analysis attempts to encompass *all* relevant facts. In human cognition, this must include data from biologically relevant sources such as hormonal and brain research and studies in nonhuman species. (1999:2, emphasis in original)

Intellectual Developments at the Turn of the Nineteenth Century

Toward the end of the nineteenth century psychology and sociology started to emerge as sciences. The progenitor of sociology, August Comte, advocated the use of evolutionary theory for the study of human society, and early psychologists, such as Wilhelm Wundt, referred to Darwin's theory as well. There was no synthesis, however. Psychologists did not study distal causes of behavior, with the one exception of William James. He believed in the importance of using the concept of adaptation in explaining the existence of certain mental features, like consciousness (Lopreato and Crippen 1999; Plotkin 1997).

This fairly promising start would soon falter. Skepticism about the strength of natural selection and about the application of evolutionary theory to humans caused Darwinian theory to lose much of its force at the turn of the century. The rise of genetics in the first decade of the twentieth century dealt a further serious blow to Darwinian ideas, because it was thought that genetics could replace Darwinism to account for evolutionary change. The conceptual marriage of evolutionary theory and genetics would only happen later, in the 1930s. At the turn of the century and into its first decade, however, Darwinism seemed to be dying (Plotkin 1997). The turn to radical environmentalism can further be understood as a reaction to the ideological use of biology made by social Darwinists. It was not safe for social scientists to embrace Darwinian theory openly, since it had been tainted (Lopreato and Crippen 1999). The evolutionary connections that had been forged by the early sociologists and by William James were thus soon lost, and would be so for a long time.

Another founder of sociology, Emile Durkheim, set the future course of the discipline in *Rules of the Sociological Method* (1895). Durkheim argued that social phenomena formed an autonomous system and could be explained only by other social phenomena. The founders of anthropology, such as Alfred Kroeber and Franz Boas, were equally united on this point. By separating psychology from biology, the social sciences withdrew from the process of scientific integration that had been unfolding since the Renaissance (Tooby and Cosmides 1992).

The Rise of Behaviorism

In psychology, behaviorism would come to dominate, especially in the United States. The birth of this school of thought, which destroyed the place of evolutionary thinking in psychology for almost half a century, is usually put at 1913 with the publication of John Watson's article "Psychology as the Behaviorist Views It." Watson rejected any explanation of behavior rooted in unobservable causes, such as mental states or consciousness, which he considered subjective and unmeasurable. Psychology had to confine itself to the study of overt behavior and how it was controlled by external stimuli. To Watson, B. F. Skinner, and the other behaviorists, the causes of behavior were not to be found inside the organism, but in the external world. They assumed that the only innate property of humans was a general ability to learn by association and by reward and punishment. This entailed that any behavior could be shaped as easily as any other merely by reinforcing it. The nature of humans, it was thought, is that they have no nature. They could be conditioned into anything.

The central tenets of behaviorism were challenged for the first time in 1959, when Noam Chomsky wrote a now-famous critical review of Skinner's work on language acquisition. In 1971 Harry Harlow's experiments with monkeys showed that baby monkeys raised in isolation preferred the comfort of an artificial terry cloth "mother" to that of a wire mesh mother, although it was the latter one that dispersed food to them. It became clear that the external environment is not the only determinant of behavior. These and other experiments eventually led to the decline of behaviorism (Buss 1999; Pinker 2002; Plotkin 1997).

Cultural Anthropology: To Tropical Paradises and Back Again

The dominating assumption in psychology and sociology that humans have no nature seemed to be confirmed by the astonishing cross-cultural variability found by anthropologists. The findings seemed to show that the human mind is indeed a blank slate, almost infinitely malleable, with all psychological phenomena, such as emotions, being social constructions. Perhaps most influential was Margaret Mead, a student of Franz Boas. To most Boasians anthropology and politics were inextricably entwined. If they were able to find cultures free from competition, rape, and murder, then perhaps people would recognize what the Boasians themselves believed to be the case: that our own problems all derived from the evils of capitalism, Western sexual morality, and Western values. According to them, Western capitalism did not write anything positive on the blank slate of the human mind.

In her 1928 bestseller *Coming of Age in Samoa* Mead purported to have discovered a society where stress, competition, problems of puberty, violence, war, rape, sexual jealousy, or sexual proprietariness were unknown. According to her, the Samoans derived their emotional well-being from their living in communes. In *Sex and Temperament in Three Primitive Societies* (1935) she described cultures in which she had found the Western sex roles to be completely reversed. With the Chambri, she said, women were dominant, whereas men were emotionally dependent. The men wore makeup, they painted and danced. With the Arapesh, the men were peaceful because Arapesh children were educated in a very affectionate way. In a similar vein Ruth Benedict, the second most important Boasian, argued in *Patterns of Culture* (1934) that culture could mold personality in any arbitrary way.

Subsequent researchers found, however, that many of the original reports of these tropical cultures were simply products of wishful thinking. Arapesh men turned out to be violent headhunters, who were dominating their wives. The face paint of Chambri men was not feminine makeup, as Mead supposed, but a sign of honor signaling that they had killed an opponent. The painting of totem poles and the ritual dances in order to inaugurate them were important male activities, with no women allowed. Anthropologist Derek Freeman found Samoa to be a very authoritarian, hierarchical, and patriarchal society, with murder and rape rates higher than those in the United States. Furthermore, the men were intensely sexually jealous. When Freeman presented the results of his forty-year study of Samoa in *Margaret Mead and Samoa: The Making and Unmaking of an Anthropological Myth* (1983), this created a storm of controversy. He was widely criticized by a social science community that had eagerly embraced the findings of cultural anthropologists such as Mead. Subsequent research has, however, confirmed Freeman's findings and, more important, the existence of numerous human universals, such as romantic love, facial expressions, and male sexual jealousy (Brown 1991; Buss 1999; Evans 2001; Roele 2000; Torrey 1992).

How is it possible that Mead could have been so misled? For one, due to a lack of time prior to her departure, she did not know the language; Freeman did. She was at Samoa for nine months, staying with white people in a comfortable house; Freeman's visits to the island amounted in total to six years, and he lived among the Samoans themselves. Mead did not consult the then available archives about life at Samoa since the beginning of the nineteenth century. Her informants were adolescent girls, who stated afterward that they had been pulling her leg. Moreover, in her fieldwork Mead was not known for her steady patience, bullying her informants and interpreters. The scientific level of Mead's work and of that of the Boasians was, in short, abominable, and at the end of the sixties it became clear that they had failed collectively. Meanwhile, however, their ideas on human nature, politics, and sexual norms were embraced by the students of that time. Boasian ideas on the nearly limitless malleability of human nature would live on for a while, not because they were scientifically based, but because of their political attractiveness (Brown 1991; Torrey 1992).

That this attractiveness endures is proven by philosopher Val Dusek, who indignantly attacks the "campaign against anthropological relativism" (1998:3) by pointing to Boas's and Mead's antiracist views and by trying to discredit Freeman as someone who "had many axes to grind and was as committed to sexual repression (including being outraged by the display of human genitals on statues in a public park) as Mead was committed to sexual freedom" (1998:3). Dusek does not mention the confirmation of Freeman's findings by others, nor Mead's obvious scientific shortcomings.

Social Constructionism, Environmentalism, and the Left

The cultural relativist position seemed to be on the brink of annihilation when its scientific inadequacies were laid bare in the 1960s. However, another new development would come to the aid of cultural relativism: epistemological social constructionism. The view that there is no objective truth and that the scientific method is just a Western, white, male product intended to dominate everything that is 'other' would become very useful for those who saw themselves confronted with scientific findings they were not willing to accept, such as findings on a universal human nature. It is no surprise that environmentalism and epistemological social constructionism often go together, as they tend to do in many strands of feminism.

How did this view come about? According to Stephen Cole (1996) and Ullica Segerstråle (2000), the single most important influence was the 1962 classic book by philosopher of science Thomas Kuhn, *The Structure of Scientific Revolutions.* Kuhn showed that the history of science is not an asymptotic curve of progress toward Truth, but that it consists of long periods of paradigmatic status quo, occasionally interrupted by shifts in the shared worldview, resulting in a new and different way of interpreting nature. The book seemed to give warrant to the view that social consensus determined 'nature' rather than nature determining social consensus—although Kuhn himself, after learning what constructionists made of his theory, rejected this interpretation. It is no surprise that feminists, too, often refer to Kuhn (Rose 1983; Rosser 1992; Segal 1999).

It has for the most part been self-identified left-wing intellectuals who have developed and embraced epistemological social constructionism.[2] This is rather astonishing. Just having a quick look at human history should make it clear that science has dispelled a lot of myths about female nature, race, sexual orientation, and many other human attributes that have long been a basis for discrimination. These developments took place in spite of science's faults. In contrast, it is difficult to see how social constructionism could offer criteria for distinguishing reliable knowledge about our fellow human and nonhuman beings from superstitious beliefs.

Moreover, as Ullica Segerstråle (2000) argues, the link between environmentalism and progressive politics, although seemingly obvious to many, is a historical contingency, not a logical necessity. After all, hereditarian explanations prevailed in academia at the beginning of the twentieth century, and they were used for both conservative and progressive social policy recommendations. She also points out that most leading evolutionary theorists are far from being politically right-winged, thereby citing sociobiologist Pierre van den Berghe, who has said that a review of leading sociobiologists would lend more credence to the contention that sociobiology is a *communist* conspiracy. John Haldane, who is generally credited for having first touched on the notion of kin selection, a theoretical cornerstone of sociobiological theory, was a leading member of the British Communist Party. So was John Maynard Smith. Robert Trivers has been heavily involved in radical black politics, and most other sociobiologists are left-of-center liberals or social democrats.

Still, many eugenicists connected their belief that genes are the crucial determinants of human behavior to a belief in the inequality of mankind. This included the proclaimed inferiority of some ethnic groups, such as Jews and peoples from Eastern Europe, as well as antidemocratic tendencies. Therefore the shift in the 1930s from hereditarianism toward environmentalism in the explanation of social behavior had scientific as well as ideological reasons. Scientifically, there was the (sloppy) work of anthropologist Franz Boas and his students, who promoted the idea of culture over biology. Ideologically, there was the dwindling support for the eugenics movement after reports of escalating Nazi sterilization practices. Scientific racism lost its respectability (Torrey 1992). These developments led to a dramatic decrease in articles on race and sex differences during the 1930s and 1940s. The final turning point came with the explicit declaration of environmentalism as the politically and intellectually correct approach by UNESCO in 1952. To secure the goal of dissociating science from racism after the horrors of the Second World War, the newly formed UNESCO had commissioned a statement on race by a group composed primarily of Boasian cultural anthropologists. This statement, published in 1950, rejected the idea of race as a scientific concept. It was subsequently reformulated by geneticists, but the argument that race was not a scientific concept remained constant (Rose and Rose 2000b). With the climate in society and academia shifting in favor of environmental explanations, they were taken for granted by a new generation of scholars.

As for social constructionism, Gross and Levitt (1994) offer several possible reasons for the frequent link of this paradigm with a leftist political position. First, it is difficult for opponents of capitalism to leave science exempt from what they regard as capitalism's omnivorous tendencies. Science is, after all, well integrated into the capitalist system. However, as Gross and Levitt object, a more correct view would be that scientists have never been granted as much autonomy of thought and freedom of ideas as today. Second, it cannot be denied that Western science has been plagued by misogyny, racism, class snobbery, and other ideological evils. As the third and most important factor lying at the social and ideological roots of antipathy toward science and its standards of validity, they see the disillusionment with post-Enlightenment humanism. In this they follow George Steiner in his 1971 essay *In Bluebeard's Castle*. Whatever may be said in praise of Western civilization, Steiner ponders, its level of cruelty and selfishness has not fallen very much below that in other times and places.[3] In the meantime our weapons have grown unimaginably lethal. Science is sometimes the eager accomplice of atrocities. Thus it should not surprise us that it becomes "an irresistable target for those Western intellectuals whose sense of their own heritage has become an intolerable moral burden," Gross and Levitt assert (1994:220). They point out something crucial:

> We are bound to Enlightenment values—the universality of moral principles, the sanctity of individual volition, a detestation of wanton cruelty—and yet we have no choice but to indict the very civilization that begat those values as it goes careening through time leaving pain, death, bewilderment, the wreckage of aboriginal tribes and of rain forests in its wake. *But again, the terms of that indictment can be spelled out only in the language of those values.* This, and not the mincing word games of the deconstructionists, is the true aporia. The criminal is also accuser and judge. (Gross and Levitt 1994:218, emphasis added)

It may thus not be possible for us to escape Enlightenment values, just as we cannot escape traditional standards of rationality and logics. Gross and Levitt further link the attempt of many contemporary leftist academics to do just this to the counterculture of the sixties, with its inclination to regard heterodoxy, per se, as intrinsically valuable. A predilection for the unconventional almost always reigns among rebellious spirits, they argue, spying a strong echo of the spirit of the sixties in modern-day challenges to science. Many academics who are most actively hostile toward science are affiliated with areas of study that first arose during that time, such as women's studies and ethnic studies.[4] In their very origins, however, these subjects were linked as much to the oppositional culture of the sixties as to the formal traditions of research and scholarship. The current phenomenon "has entirely lost its playfulness and now parades as serious scholarship and solemn theory" (Gross and Levitt 1994:223).

Everything we have seen thus far seems to lead us to the conclusion that (1) it is impossible to leave traditional standards of epistemology and ethics *completely* behind, and (2) that people who aspire to do so are basically trying to find convoluted ways of making sure that they reach their political ends, *while this move is not necessary at all in order to get that far.* One need not assume that science is just a contingent social construction in order to condemn its sometimes disastrous effects, or in order to make sure that non-Western peoples' habits and belief systems are respected—that is, inasmuch as these habits and beliefs are not in flagrant contradiction with basic human rights. One need not assume that women and men are the same in order to make sure that they get equal opportunity and are equally treated before the law (although, indeed, it will be a matter of debate what "equal" in this case means—does it just mean offering them the same opportunities, or does it imply adjusting for women-specific elements such as pregnancy?).

So we have gone from a fairly promising interdisciplinary start of sociology and psychology at the beginning of the twentieth century to the intellectual isolationism that has become politically correct since the 1950s. Since that time, however, many new developments have laid bare the shortcomings of environmentalism. While we will have a look at these in the next chapter, I focus first on three misconceptions that plague environmentalism: a false dichotomy between nature and nurture, the myth of genetic determinism, and the naturalistic fallacy.

The Nature-Nurture Controversy

Although the falseness and outdatedness of the dichotomy between nature and nurture has been laid out countless times,[5] the misconception stubbornly persists.

In its simplest form, human behavior is regarded as *either* biologically based *or* culturally constructed, and we have to find out which aspects of behavior are which (Donovan 2000; Nicolson 2000). In a somewhat more sophisticated vein, a neat distinction is made between nature and nurture, with nurture overtaking where nature ends. The key question then becomes: Where does nature end and where does nurture begin (Brouns 1995b)? The accusation of biological or genetic determinism frequently launched at the address of sociobiologists and evolutionary psychologists (e.g., Benton 2000; Coyne 2000; Gray 1997; H. Rose 2000; Rose and Rose 2000b; Shakespeare and Erickson 2000; Tobach and Reed 2003; Travis 2003b) testifies to this kind of (sometimes willful?) misunderstanding. When one adds the (correct) complication that our perception of nature is always guided by cultural meanings and values, we are getting at the (incorrect) postmodern conclusion that nature, too, is culturally constructed. Sex thus becomes subsumable under gender (Birke 1999; Butler 1990, 1997; Nicholson 1994).

This latter option is, indeed, one possible way of leaving the nature-nurture dichotomy behind. The problem is that the world does not seem to be constituted like that, as converging evidence from multiple scientific disciplines shows (see next chapter). Moreover, all of the aforementioned views of the relationship between nature and nurture are based on a conception of biology that is fundamentally wrong.

As Margo Wilson, Martin Daly, and Joanna Scheib (1997) point out, many people falsely equate biology with its subdisciplines, such as genetics, endocrinology, and neurology. Moreover they think of biological influences as intrinsic and irremediable. From there it is a short step to seeing these biological influences as the antitheses of extrinsic and remediable social influences.

Strictly speaking, however, biology is the study of the attributes of living things, and since only living things can be social, the study of social behavior also pertains to biology. The developmentally, experientially, and circumstantially contingent variations manifested in behavior are inherently part of a biological mode of explanation. Thus every aspect of any living thing is, *by def-*

inition, biological.[6] Furthermore everything biological is always a result of interaction between genes and environmental factors. Even an individual cell is a product of genes and the environment (e.g., various chemicals). And since all traits and behaviors require environmental input, this means that they can be altered by manipulating one or more of their developmental causes.

The nature-nurture dichotomy as it is often found in the writings of feminists and social scientists dates from the 1950s, when ethologists were defending adaptive behavior to be "instinctive," whereas psychologists were stressing learning and nurture. Current scientific theorizing on the subject, however, has become much more sophisticated, with the breach between nature and nurture having been filled for the first time by ethologist Konrad Lorenz in his 1965 book *Evolution and Modification of Behavior.* Lorenz's great insight was that learning *itself* is adaptive. Genetic information somehow ensures that only certain events or acts or relationships are learned. It gives rise to neural structures that are a consequence of past selection pressures. They make sure that the individuals of a species will filter out the irrelevant information and learn what they need to learn in order to survive and reproduce (Plotkin 1997).

The next chapter describes the accumulating evidence from many different disciplines that learning is indeed constrained. No learner of any species is a generalist. Thus the common assertion made by feminists and social scientists that cultural learning is not biological is wrong. Learning is, as are all other traits and behaviors, a product of gene-environment interactions. We easily learn to become afraid of heights and spiders, but not of cars and electrical outlets. The latter have been invented too recently. Although they constitute real hazards for us, there has not been sufficient time to evolve psychological adaptations against them.

It is a crucial observation that these learning constraints are caused by past selection pressures. Thus, as Plotkin, and many others with him, concludes: "Nature and nurture are inextricably enfolded within one another because nurture has nature, and yet nature must be nurtured and nurture is a part-cause of nature" (1997:68–69).

The same applies for culture. One cannot contend that some human behaviors are biological and others are cultural and hence nonbiological. Some *differences* in behavior between individuals could indeed be caused entirely by cultural influences. That does not mean, however, that an individual's culturally influenced behavior is entirely caused by the environment. Cultural behavior is always a product of gene-environment interactions. And we can learn nothing without underlying adaptations for learning.

A good example of this is language. In order to speak a language, social learning is needed, so it is clearly a cultural trait. One cannot learn a language, however, without specialized underlying brain structures. These structures (modules) ultimately evolved through natural selection, but are at the same time the proximate product of complex interactions between genes and environment during a person's life. Hence, although language is cultural, it is just as much biological (Pinker and Bloom 1992; Pinker 1997). If we want to understand any particular fact about humans, we need to analyze how their cognitive architecture, which is a product of evolution and is embodied in a physiological system, interacts with the world around it.

Do these insights on the nature of biology testify, as psychologist Lynne Segal insists they do, to "a biological absolutism more incompatible than ever with any serious recognition of the dynamics of culture, and its role in the formation of human existence" (1999:99)? Segal is quite sure that the objections of those who stress the significance of culture do not derive from "misinformation" or from "a refusal to listen," but from "obstinate attention to certain new twists in the curtailing of culture" (1999:99). What we find here, however, does not really fit the picture of biological absolutism as she paints it. The problem seems to be that she fails to grasp the multilayeredness of the model. She interprets the interactionism as a reductionism, in the sense that she thinks that an evolutionary framework leaves no room for environmental or cultural influences. The irony is, however, that evolutionary theory is *as much* about the role of the environment in ontogeny (the development of the individual) and phylogeny (the evolution of the species) as it is about biologically based dispositions. Adaptations can only develop when confronted with the right environmental triggers, and a change in the environment will in the long run cause evolutionary change.

The Myth of Genetic Determinism

No one contends that genes alone direct development and behavior. It is often the failure to understand evolutionary psychology's interactionism that leads many feminists to the wrongheaded accusations of genetic determinism and biological determinism (with regard to the latter it must be repeated that biology refers to any aspect of living organisms, which means that the phrase 'biological determinism' is meaningless). Whereas critics equate evolutionary mechanisms with behavioral inflexibility, a fundamental message of evolutionary psychology is that evolution sculpts organisms to behave contingently on their environment and their own current state.

This view is consistent with what modern genetics tells us about the workings of genes. Genes are designed to take their cues from the environment. They remain active during life, responding to experience, switching on and off, allowing us to learn, to remember, to imitate, and to absorb culture. They are the agents of nurture, predisposing us to extract certain kinds of information from the environment. They push us toward the goals that are physically and emotionally rewarding to us, such as food, social status, and affection—things likely to have had adaptive value in our evolutionary past. Small genetic differences between people will be multiplied by the environment. Sporty children will want to practice sports, intellectual children will want to read, and the more time they spend at doing what they enjoy, the better they will become at those tasks. They create their own environments, as we all do, seeking out those stimuli that are rewarding to the evolved beings and unique individuals that we are. The way we live our lives will at the same time affect which genes express themselves (Jacob 1997; Ridley 2003).

Some critics, such as biologist Sue Rosser, continue to represent current evolutionary theories in a highly misleading way: "Emphasizing separation or differences between males and females often leads to a form of biological determinism implying that differences in social and cultural role, status, and behavior are caused by genetic, hormonal, or anatomical differences between the sexes" (1992:157). By crudely distorting the subtlety of the interactionist model, Rosser plays along with the common misconception within academic feminism that evolutionary psychologists leave no room for environmental influences. Her political motivation for doing this is evident from her argument: She is afraid that differences in social status between the sexes will be accepted as natural and inevitable.

However, "[n]either 'biology,' 'evolution,' 'society,' or 'the environment' directly impose behavioral outcomes, without an immensely long and intricate intervening chain of causation involving interactions with an entire configuration of other causal elements" (Tooby and Cosmides 1992:39). At each link of such a chain intervention is possible, so that we might change the final outcome. Talking of genes is, furthermore, not even *needed* in evolutionary theorizing. Although most biologists do assume that genetic mechanisms of heredity are common and important, all they need to know is that phenotypes vary and that there are correlations between phenotypes and survival and reproductive success. The exact mechanisms of heredity do not have to be clear in order to study evolution—Darwin did not know them either (Gowaty 1995; Waage and Gowaty 1997).

Apart from that, one might wonder why genetic influences are thought to be so much more inescapable in their effects than environmental ones. As Richard Dawkins points out, genetic and environmental causes are in principle no different from each other. Some may be hard to reverse; others may not be so. A child whose mathematical deficiency has a genetic origin may flourish when it gets the right teacher, for instance. Severe parental neglect during childhood may, however, have ineluctable negative consequences on the later emotional development of a child (Radcliffe Richards 2000).

Moreover, if one is shy because of something that happened to one as a child, this is no less a determinist event than if one's shyness had a genetic basis. We are all the product of the many determinants that have shaped us to the persons we are now. The real mistake is to equate determinism—be it environmental, genetic, or some other kind—with inevitability. Determinism is about the causes of a particular situation, not about the consequences. Having a strong inclination or emotion, regardless of its origins, never forces people to act in a way determined by it (Ridley 1999). And even if it is true that characteristics are mainly socially induced, we would still need to know how to control them. Experiments in changing human nature by large-scale political reorganization have not yet shown much sign of achieving their purpose.

The Naturalistic Fallacy

It may still not be clear *why* exactly the idea of a nearly infinitely malleable human nature is politically attractive to some. One might object, for instance, that on these conditions it would be relatively easy to install a totalitarian political regime (just make sure that humans are molded to obey authority, no matter how that conflicts with their own interests). Moreover, and worse, there would be no objective basis for contesting this state of affairs. Human nature is a yardstick for measuring evil. If there is no human nature, then there is nothing wrong with, for example, teaching women that it is a pleasant experience to be raped, beaten, and oppressed. The sheer outrageousness of this proposal already indicates that a gross inconsistency is lurking within the heart of a radical environmentalism.

As we have seen, however, it is not surprising that when the idea of environmentally produced differences between the sexes came up, feminists were eager to embrace it, since women's oppression had long been justified by means of biological arguments. At the time the enthusiastic feminist reception of works such as Simone de Beauvoir's *The Second Sex* (1949), which explicitly defends an environmentalist view, may have been a very understandable move. One would expect this vigorous reaction against previous misconceptions of the nature and the appropriate place of women in society to boil down after a while, however. It should have made place for a more balanced view of the sexes, which takes into account all available evidence on the ways men and women are constituted. It did not work out that way. On the contrary, the environmentalist position seems to have become only more extreme with time.

Androcentric bias and a lack of theoretical sophistication within early sociobiology have probably scared many feminists away, something that is, again, quite understandable. When we see sociobiologist David Barash equating femininity with noncompetitiveness (1979:113), for instance, an alarm bell starts ringing. The female-friendliness and explanatory power of current evolutionary theorizing should, however, be able to appeal to feminists. It does not. Doreen Kimura points out another possible reason for this state of affairs:

> The bias against biological explanation seems to have arisen from egalitarian ideologies that confuse the Western concept of equal treatment before the law—the societal application of the idea that "all men are created equal"—with the claim that all people are in fact equal. People are not born equal in strength, health, temperament, or intelligence. This is simply a fact of life no sensible person can deny. We have chosen a system of governance which has decided that *despite such inequalities* each individual shall have an equal right to just treatment before the law, as well as equal opportunity. (1999:3, emphasis in original)

The confusion between both kinds of concepts is an example of the naturalistic fallacy. According to Oliver Curry (2002), there is not one 'naturalistic fallacy' (NF), but there are seven: (1) the direction of evolution is the direction that we ought to go in; (2) what currently exists ought to exist; (3) natural is good; (4) good is identical with its object (i.e., it is possible to exhaustively define good in terms of properties such as pain and pleasure—the NF as defined by George Moore); (5) good is a natural property; (6) you can go from facts to values; (7) you can go from "is" to "ought" (the NF as defined by David Hume).

The naturalistic fallacy as encountered in feminist theory is usually a mix of (3), (6), and (7): What ought to be is defined by what is, and especially by what is natural. Margo Brouns commits it when she writes that "any statement about how women 'are' is at the same time an implicit statement about what they 'ought to do'" (1995b:44, my translation). On the next page, Brouns unwittingly contradicts herself when noticing that the views of the sexes as fundamentally the same and as fundamentally different can both be used to justify the struggle for equal rights. The link between "is" and "ought" turns out to be not so stringent after all.

This lack of compelling reasons for going from is to ought, or for arguing that what is natural is good, should be obvious when we think of disease, death in childbirth, or natural disasters. For tens of thousands of years, people have used their "scientific" insights to fight the harsh environment they found themselves living in: By making fire they could protect themselves against the cold and against predators; using weapons turned out to be a much more efficient way of hunting. One of the reasons why we have developed science and technology is precisely that we wanted to overcome our natural vulnerabilities. Sure, (pseudo) scientific insights have sometimes been used for political reasons, but therefore to call science "an ideology" and "the great legitimator" (Rose 1983:274) seems shockingly simplistic. Despite its obvious shortcomings as a human endeavor science is, in the first place, a quest for the ways the world works. Then why does Dorothy Nelkin describe an evolutionary approach to human nature as "a guide to moral behavior and policy agendas" (2000:20) and why does Zuleyma Tang-Martinez write that it "serves only to justify and promote the oppression of women" (1997:117)? Why have biologist Randy Thornhill and anthropologist Craig Palmer been accused of being ideological, of justifying rape, and of blaming the victim after the publication of their 2000 book *A Natural History of Rape: Biological Bases of Sexual Coercion*, not only in countless reviews in the popular press, but also by serious academics (e.g., Kimmel 2003; Shields and Steinke 2003; Rosser 2003; Wertheim 2000)?[7]

According to Thornhill and Palmer themselves (2000), one reason is that the naturalistic fallacy is often committed in the writings of social scientists themselves, and that as a result social scientists assume evolutionary theories as well to be intended to imply a position about how the world ought to be. That might well be one explanatory element. Apart from that, it is, as I pointed out earlier, quite "reasonable" for those theorists who, like Fausto-Sterling, think that "*all* scientific writing embodies political agendas" (1995:171, emphasis in original) to infer that evolutionary scientists are making political statements. The thing is that feminists are not always chasing phantoms. It cannot be denied that some dubious things have indeed been asserted, something that evolutionists seldom acknowledge in their many refutations of the naturalistic fallacy. Sarah Hrdy (1999a) notes that as late as 1980, one can find it seriously suggested that differences in competitiveness between girl and boy athletes may be adaptive for our species and therefore should not be erased. Whereas this is an instance of the naturalistic fallacy, the following quote is an expression of genetic determinism: "Men and women differ in their hormonal systems . . . every society demonstrates patriarchy, male dominance and male attainment. The thesis put forth here is that the hormonal renders the social inevitable" (Steven Goldberg 1973, as cited in Fausto-Sterling 1992:91).

Both fallacies are closely entertwined, since genetic determinism amounts to nearly the same as saying that the behavior *should* exist, because it is the way nature works.

The observation that there has been flawed reasoning does not, however, warrant the fabrication of quotations from others, as Hilary Rose does. Rose (2000:116) cites sociobiologist David Barash as having written "If Nature is sexist don't blame her sons" (in *Sociobiology: The Whispering Within*, 1979:55). The phrase does not appear anywhere in Barash's book. On the contrary, he points out more than once that the fact that some behavior is natural does not mean that it is good or that it ought to be. The misunderstanding seems to have arisen from this passage:

> To my thinking, sexism occurs when society differentially values one sex above another, providing extra opportunities for one (usually the males) and denying equal opportunities for the other (usually the females). As such, it has nothing to do with sociobiology. On the other hand, sexism is also sometimes applied to the simple identification of male-female differences, and on this count, sociobiology is, I suppose, sexist. No one would think it awful to state that a man has a penis and a woman, a vagina. Or that a man produces sperm and a woman, eggs. But when we begin exploring the behavioral implications of these facts somebody is sure to cry "Foul." If male-female differences are sexist, we should put the blame where it really belongs, on the greatest sexist of all: "Mother" Nature! (Barash 1979:90)[8]

Rose may also have conflated her reading of Barash with a misrepresentation in an article by Ruth Hubbard:[9] "As sociobiologist David Barash presents it, 'mother nature is sexist,' so don't blame her human sons" (Hubbard 1988:8). Needless to say, such distortions of the original material, by Rose and by Hubbard, do not attest to good scholarship.

My examples of the naturalistic fallacy and of genetic determinism date from a few decades ago, and erroneous reasoning like this is not representative at all of current scholarly writing. I suppose everyone will agree that it would be unfair to judge current evolutionary theorizing on humans by mistakes made in the past or by the (inevitable?) straying of some scholars. Yet these examples, certainly when seen together with the androcentric bias obvious in much of early sociobiological views of the sexes, may also help to explain the feminist wariness of putative instances of political intentions within current evolutionary theorizing. Many prominent critics (e.g., Fausto-Sterling, Hilary Rose, Segal) seem to have been so put off by early sociobiology that, if they concern themselves with the current evolutionary approach to humans at all, they find it impossible to evaluate it in a relatively unbiased manner.

The harsh criticisms, (Darwinian) feminist as well as nonfeminist, of early sociobiology have helped the field to mature, to become more sophisticated and more scientifically based. But for many feminists the case seems to be closed. In 1997 Fausto-Sterling made an update of her thinking on sociobiology. Her list of references, however, hardly contains any publications of the 1990s. Yet she feels quite confident to stick to her argument that "[b]iological claims about social difference are scientifically invalid" (1997:58). In her contribution to *Alas, Poor Darwin*, "Beyond Difference: Feminism and Evolutionary Biology" (2000a), she does not do any better.

As for the naturalistic fallacy, there is, indeed, the danger that evolutionary insights might be used to justify a conservative political agenda. Understandable as the feminist fear of naturalizing and legitimizing traditional gender roles may be, one would think that feminists would better devote their time and energy to informing the public about the predominating misconceptions regarding biology, instead of reinforcing them by writing things like "[e]volutionary principles imply genetic destiny" (Nelkin 2000:22).

Second, there is the danger that criminals might be exonerated for their behavior on the basis of evolutionary arguments (Fausto-Sterling 1992). But, as Anne Campbell (2002) argues, the misuse of scientific research by defense attorneys, being doubtless widespread, is certainly not confined to evolutionary theory. Not only is contemporary work on neuroanatomy, hormones, and clinical disorders misapplied in this way, environmental factors are far more commonly misused. Parental abuse and neglect, inadequate educational opportunities, and drug addiction have all been put forward to mitigate guilt and reduce sentences. The issue here, as she rightly stresses, is not evolutionary theory but the prevailing legal philosophy of responsibility and free will. Moreover any cause of behavior, not just the genes, raises the question of free will and responsibility. "The difference between explaining behavior and excusing it is an ancient theme of moral reasoning, captured in the saw 'To understand is not to forgive'" (Pinker 1997:53).

Besides, as legal scholar Owen Jones (2001) writes, evolutionary theories address statistical probabilities and predictions. They cannot explain in detail why this particular individual, in this particular circumstance, did what he or she did. Jones deems it highly uncertain that criminals will ever be exonerated on the basis of evolutionary arguments.

Current Socialization Theories

Feminists typically explain gender differences through theories of socialization.[10] No single researcher will deny that socialization practices are indeed important in channeling male and female behavior. Behavioral ecologist Bobbi Low (1989), for instance, has shown that the degree to which boys and girls are trained differently cross-culturally depends in a predictable way on the kind of society they live in: stratified or nonstratified, polygynous or monogamous, patrilocal or matrilocal.

Low's evolutionarily based approach, which assumes an interaction between genotypes and environments, differs tremendously from the dichotomous way in which sex differences are explained in traditional feminist accounts. The latter often seem inadequately thought through. Not only do they assume the human mind to be more or less a blank slate—a concept dating back from behaviorist times—they also seem to regard it as an entity floating free from the body. They retain the Cartesian distinction between the body as a physical organism and the mind as nonphysical. The evidence is overwhelming, however, that our mental life depends entirely on patterns of physiological activity in our brain, the basic plan of which is largely shaped in the womb. People born with variations on the typical plan have variations in the way their minds work. Although learning and practice certainly affect brain structure—indeed, *consist of* implementing changes in connections between brain cells—most neuroscientists believe that the brain is not indefinitely malleable by experience (Pinker 2002). Angier (1999) criticizes her fellow feminists for rejecting all arguments having to do with hormones, with the senses, or with the body. They consider the body to be only a vehicle, never the operator, as if humans were *completely* different from other species, she says. In this respect I can only agree with her.

Rosser, for instance, accounts for sex differences in the domain of science in this way: "The forces of socialization begin at birth and continue throughout an entire lifetime." Although "[e]ach educational or socialization incident may be minor," added together they may be the "final factor" that steers women away from decision-making positions in science (1992:36–37). Antonia Abbey and her colleagues provide us with one of the typical feminist theories of socialization:

> From birth, girls and boys are frequently treated differently by parents, peers, and teachers, thus shaping their responses in gender-specific ways. Through direct reinforcement and vicarious learning through modelling, members of a culture learn to behave in ways viewed as appropriate for their gender and to internalize gender-consistent self-schemas. (Abbey et al. 1996:139)

Or consider Marilyn French:

> [W]ar requires fighters, and people who have not been indoctrinated into a gender cult, have not been taught that aggression equals identity, do not want to fight. To get men to fight rather than flee, male leaders had to turn them against life, identified with women, sensual pleasure, children, the growing and eating of food. (1992:179)

Pornography and the mass media are often mentioned as important ways of shaping and perpetuating gender roles (Brownmiller 1975; Denmark and Friedman 1985; Dworkin 1997; French 1992; Michielsens et al. 1999; Schwendinger and Schwendinger 1985). Sometimes vague references are made to "patriarchy," the "gender divisions of labour," and the "practices that shape and realize [heterosexual] desire" (Connell 1995:46–47). Some go so far as to claim that heterosexual desire is "eroticised power difference," with no roots in biology at all (Jeffreys 1990, as cited in Patai 1998:130). Some, realizing that sex-specific socialization alone cannot possibly explain the ubiquitousness of the same gendered patterns throughout the world, point to psychoanalytic theory to further account for this phenomenon or to account for the frequent failure of the molding of gender identities (Chodorow 1978; de Beauvoir 1949; Gilligan 1982; Segal 1999; ten Dam and Volman 1995).

Some do refer to the body, but in the way mentioned earlier: with the mind floating free from it. Psychologists Alice Eagly and Wendy Wood, for instance, point to men's greater size and strength and to women's childbearing and suckling ability, physical differences that, according to them, "interact with shared cultural beliefs, social organization, and the demands of the economy to influence the role assignments that constitute the sexual division of labor within a society and produce psychological sex differences" (1999:409). While they recognize those physical differences to be the result of evolutionary pressures, they do not seem to recognize them as *automatically implying* behavioral and psychological differences. Physical characteristics do not evolve out of the blue. Our upright posture implies bipedalism. Our digestive system reflects what our ancestors ate. Functional breasts would not have evolved without the simultaneous evolution of particular behavior patterns, such as placing an infant to the breast.[11] Greater upper-body muscle mass in males would not have developed without the simultaneous evolution of behavior such as punching and grabbing. Moreover, the evolution of these behavior patterns implies psychological adaptations, both cognitive and emotional, to guide those behaviors. Without inherent goals and drives an organism would not do anything (Thornhill and Palmer 2000).

References to patriarchy and to gender divisions of labor are merely descriptions of existing phenomena; they do not *explain* anything. Psychoanalytic theories cannot help us any further either if we want our explanations of sex differences to be firmly scientifically grounded (Crews 1996, 1998; Grünbaum 2002; Israëls 1999; Torrey 1992). The other above-mentioned socialization explanations depend heavily on the concept of a blank slate mind being molded by reinforcement (behaviorism) and imitation (social learning theory). Nobody doubts that reinforcement can indeed shape behavior and that imitation is a central primate capacity. The question is, however, whether these effects are strong enough to account for the patterns of sex difference that are noticed worldwide.

Research from a variety of disciplines, including the biological and evolutionary sciences, indicates that the bottom line of the existence of sex differences lies elsewhere. Yet socialization explanations remain predominant in feminist literature as exhaustive accounts of the phenomenon. They represent what Tooby and Cosmides (1992) have coined the Standard Social Science Model (SSSM), which has served as the intellectual framework for the organization of the social sciences for a century. In a simplified form, this model holds that whatever innate equipment infants are born with must be highly rudimentary, such as an unorganized set of drives, plus the ability to learn. Mental organization is acquired from some source outside themselves: the social and cultural world. This sociocultural level is an autonomous and self-caused realm. The SSSM denies that human nature can play any notable role as a generator of significant organization in human life. It is acknowledged to be a necessary condition for it, but merely as embodying the capacity for culture. Thus general-purpose learning becomes a central concept.

Evaluating Some Environmentalist Contentions

Critics of research on the biological basis of sex differences typically nitpick at each study. They demand much higher standards of evidence than they would for socialization theories, offer alternative and sometimes far-fetched explanations, and do not take into account well-established, corroborating evidence from other disciplines.

Fausto-Sterling's *Myths of Gender* (1992) is such a critical study, serving as an important work of reference for many feminists. While I think—as doubtless many scientists do with me—that sternly scrutinizing existing research is essential to good science, Fausto-Sterling's work is, unfortunately, not always to be trusted—something to which her politicized stance might have greatly contributed. Her critiques of sociobiology and evolutionary psychology, for instance (Fausto-Sterling 1992, 1997, 2000a), consist virtually solely of political imputations, misunderstandings, selective reading, and misrepresentations and are characterized by a lack of knowledge of the main principles of evolutionary biology and of the theoretical developments of the 1990s. To give just two examples: She does not seem to understand the tremendous importance of the ultimate-proximate distinction in evolutionary biology (see next chapter for an explanation of this distinction). She considers it a trick, "clever, because it is totally unassailable" (Fausto-Sterling 1992:193). In her contribution to *Alas, Poor Darwin* (2000a), she seems ignorant of the evolutionarily inspired and confirmed hypothesis about superior female spatial location memory, a finding that is mentioned in all major evolutionary textbooks of the 1990s. Rather embarrassingly, her ignorance about evolutionary psychology leads her to hypothesize that females might well have evolved well-developed spatial and memory skills, adding that "without more data and a far more specific hypothesis, we have no way of knowing" (Fausto-Sterling 2000a:177–178). Because she is regarded as an authority in the field, however, many feminists feel perfectly happy putting their trust in her work and hence denouncing evolutionary accounts of gender as obviously flawed and politically inspired.

The list of possible remarks on her writings is long; I will offer only a few more examples here.

In her attempt to disprove the 'naturalness' of the sex gap in physical strength she writes that, whereas in the 1976 Olympics female swimmers were 10 percent slower than men for the 100- and 400-meter freestyle, now "East German female swimmers . . . swim a mere 3 percent more slowly than the men" (1992:220). Thereby she conveniently ignores that their prowess was not only due to training, but also to steroids (Young 1999).

When claiming (Fausto-Sterling 1992) that studies on mathematical achievement of the sexes done before 1974 found significant differences, but that the difference today has been reduced to almost zero, she fails to mention that one of the reasons is that out of political correctness items performed better by one sex have been deleted from the tests (Kimura 1999).

When arguing (Fausto-Sterling 1993) that the very idea of two sexes is a cultural construct since there is a large number of intersexed children, her only proof is a 4 percent figure coming from an informal estimate by gender expert John Money, who has turned out to be rather unreliable. In *Sexing the Body* (2000b) her frequency estimate of intersex conditions has dropped to 1.7 percent, a number that she has found by leafing through the medical literature. As pointed out by Leonard Sax (2002), however, her definition of what counts as intersex is too broad. Clinically, intersex conditions are those in which chromosomal sex is inconsistent with phenotypic sex, or in which the phenotype is not classifiable as either male or female. Fausto-Sterling defines as intersex, however, all those individuals who deviate from the Platonic ideal of physical dimorphism at the chromosomal, genital, gonadal, or hormonal levels. Thus she includes, for instance, individuals born with chromosomal variations other than XX or XY, all of whom develop as phenotypically normal males or females. Misleadingly, all the case histories that she presents in her book are instances of true intersexuality. When the clinical definition is applied, the prevalence of intersex drops to a mere 0.018 percent, which is almost 100 times lower than Fausto-Sterling's estimate. She has artificially inflated the prevalence of intersex conditions in light of her politics: She wants to propagate a flexible gender system. Her unabashedly political approach to scientific knowledge has led her to distort the evidence. There is, moreover, no discussion about how this view of sex as a cultural construction would mesh with theories of reproduction or with evolutionary biology.

This is not to say that there is *nothing* to be said for her point of view. Labeling harmless conditions such as intersex as pathological is indeed, as she contends, an expression of social preferences. So is the precise interpretation of male and female bodies, which to some extent differs through time and space. Whereas a muscular build in females might have been considered masculine once, for instance, this is no longer the case today. I see no benefit in denying, however, that we are *essentially* a dimorphic species, as long as it is acknowledged that there can be variations on the basic theme. We might even decide to call these variations a third sex.

Fausto-Sterling allows a role for biology in defending developmental systems theory (DST), which stresses an interactionist framework in which all levels of nature/nurture, from nerve cells to interpersonal interactions, co-produce one another. In this framework, however, talking about innate predispositions is regarded as dichotomous and genetically determinist, because according to DST there is no underlying program. The genome has no privileged role in development (Fausto-Sterling 2000b; Gray 1997). In this way a potentially limitless flexibility of behavior seems to become possible. DST cannot account, however, for the many universal patterns of human behavior and the universal differences between male and female minds (e.g., Brown 1991; Buss 1989; Schmitt et al. 2003a, 2003b, 2004); nor can it explain that the brains of girls and boys are already wired differently at birth (Kimura 1999, 2002). That Fausto-Sterling's preference for DST has more to do with dogmatism than with a scientific spirit is shown by her refusal to go into the argument that the expression of very unpopular sexualities, such as transsexualism, despite strong contrary social pressure points toward the existence of prenatally determined dispositions. Her dismissing reaction is that this form of interactionism "calls for a large dose of body and only a little sprinkling of environment" (2000b:259).

Doreen Kimura reacts to critics like Fausto-Sterling as follows:

> Such critics often use the word *prove*, suggesting that a particular study, because of some limitation (and no study of human beings can ever be perfect) does not prove a hypothesis. Quite true, no one study is likely to sufficiently support a hypothesis to the point where we accept it wholly. Scientists do not expect to prove a position, they expect either to disprove it or to find sufficient and wide-ranging evidence for it, so that it becomes more and more plausible and alternative explanations become less and less likely. Human behavioral science, especially, must operate by looking at the cumulative evidence, not just at one study. (Kimura 1999:5, emphasis in original)

Another important criterion for accepting a hypothesis, Kimura continues, is how well it fits with facts in other related fields of science, such as physiology, neuroscience, or evolutionary biology. As we see in the next chapter, the environmentalist account of sex differences does not fit in at all with other scientific fields, whereas an evolutionarily based approach does.

If environmentalism were correct as an exhaustive explanation of sex differences, the world would look completely different. First of all, we would not find this curious pattern of similar sex differences around the globe. Evolutionary psychology is sometimes charged of describing "the obvious." The obvious is only obvious to us, however, because it is part of human nature. In principle countless other ways of living are imaginable. People might not live in groups and might meet members of the other sex only for mating, for instance. Or they might as a rule find elderly people more physically attractive than younger ones. According to the environmentalist paradigm, whenever a hitherto unknown people would be found, the prediction of it being a matriarchy with women desiring younger men and men doing the bulk of the childcare would be as plausible as any other—or, for that matter, a society where no gender distinctions are made at all. It might even be a society where people are conditioned to be sexually attracted to tree trunks, as Ellis (1992) notes.

Such societies do not exist. As we saw, everywhere people live in groups, speak, have characteristic emotions, know violence, rape, and murder, have inequalities of power and prestige, are hostile to other groups, have romantic relationships, make tools, have sexual regulations, favor their close kin, make plans for the future, have a sense of rights and obligations, have rituals, know how to dance and make music, and so on (Brown 1991). Evolutionary psychology is the only theory that can explain why the human mind should show this underlying universal design. If children really were that malleable, moreover, they would not be confounding all of their parents' expectations from the moment they are born. People would be easily drilled and indoctrinated, in the way one can read it in dystopian science fiction literature.

An experiment that has come to be known as the Baby X experiment has been considered by some as definite proof that gender differences are the result of socialization. In a room with a number of toys a woman was asked to look after a baby for a few minutes. When she was told the baby was a girl, she more frequently offered it a doll to play with. As Campbell (2002) argues, however, the reason for this behavior does not seem to be what defenders of the behaviorist account, which holds that sex differences derive from boys being encouraged to fight and climb trees and girls being forced to wear dresses and play with dolls, think it is. It is quite likely that parents are simply responding to the child's own preferences, since it has become clear that children are active participants in their own development. They prefer sex-appropriate toys even when they are not specifically encouraged to do so. Sex-differentiated preferences for toys have been found in infants from nine months of age. Moreover, the Baby X finding does not prove anything unless it can be shown that the toys changed the child's subsequent behavior. If the behaviorist account were correct, we would furthermore expect the most sex-typed adults to have the most sex-typed children. No such relationship has been found.

As for imitation theory and other kinds of environmentalist accounts, children not only have their own toy preferences; they also prefer to interact with members of their own sex and show sex differences in social behavior before they know their own sex or before they can label these behaviors as being more common among boys or girls. Even at seven years of age, there is no relationship between children's gender knowledge and how sex-stereotypic their own behavior is (Campbell 2002).

Sex differences in social organization found in human children from diverse cultures parallel those found in chimpanzees, which may suggest a genetic basis for these behaviors (Benenson 2002). While this may at first seem like an unfounded claim, toy preferences—and all behaviors so far studied that are sexually dimorphic—have indeed been established to be related to the influence of sex hormones, the working of which is genetically based. In mammals the "default" form of an embryo is a female. In order to yield a male, the masculinizing influence of androgens is needed. These androgens, being released in several stages of embryonic development, not only make sure that male genital structures are produced; they also have organizational effects on the brain, with lifelong effects on behavior. In their absence the female brain pattern develops.

We know this partly from human sexual anomalies, such as cases of congenital adrenal hyperplasia (CAH). These are individuals (male and female) who, prenatally, have been overexposed to androgens. The girls are born with partially masculinized genitals and turn out to be more tomboyish than unaffected girls. Their toy preferences are strikingly similar to boys and they have a higher incidence of homosexual fantasy and preference (Kimura 1999, 2002).

Some feminists seem to think that sexual anomalies testify to the social construction of sex differences. Karin Spaink (1994) claims that transsexuals (people who typically desire to become a member of the opposite sex) are the ultimate proof of biology-as-destiny being nonsense, since according to her their gender identity is not determined by their biology. Transsexuals may, however, be the ultimate proof that it is radical environmentalism that is nonsense (the biology-as-destiny way of thinking is erroneous anyway). Transsexual people have been raised and their behavior has been reinforced according to their anatomical sex. Still they develop a gender identity that does not conform with it. An explanation may be found in Bailey (2003), who suggests that variation in prenatal exposure to sex hormones might be the basis for variation in sexual orientation. Since not all behaviors are susceptible to hormonal influences at the same time, there are several moments when variation from the "norm" can occur. This may, for instance, lead to a non- or little-masculinized brain structure in a fully masculinized body.

Another strong indication of the importance of early brain organization is provided by cases in which a boy's penis had been severely damaged shortly after birth. Because constructing a vagina is easier than constructing a penis, in the 1970s the solution in such rare cases used to be to surgically alter the boy and to raise him as a girl, a practice that had been installed by John Money, a then-renowned expert in gender identity. The first and most famous case of such a gender reassignment is that of David Reimer. At the time known as the John/Joan case, it was readily embraced by the feminist movement and the social sciences as evidence that newborns are psychosexual blank slates. It has, however, turned out to be a failure (Colapinto 2000).

John Money knew the operation to be a failure from the outset. He has, however, never acknowledged this, not even when the tormented girl—who never had shown any feminine behavior and even tried to stand up to urinate—in 1981 let herself be made into a boy again, at the age of seventeen. By the age of eight, she had already felt strongly that she was a boy. When at fourteen she decided to live as a male, her parents had finally decided to tell her the truth.[12]

When the failure was brought out in the open in 1997, many intersex people who had undergone normalizing genital surgeries as babies (as Money had advised) now came to speak out about their lives. Their stories were similar to that of David Reimer, in that being born intersexed, they had a complicated sexuality and sense of self. The point is that many of them had never been able to feel that the sex to which they had been surgically reassigned fitted them. They are now still lobbying for the abolishment of genital surgery on infants until these are old enough to speak for themselves (Colapinto 2000; Fausto-Sterling 2000b). They can then choose the sex they want or, indeed, choose not to have surgery at all. There is no reason why society should force intersexed people into a binary scheme, instead of plainly appreciating them as they are.

One criticism of the Reimer case is that the infant boy lost his penis at age eight months, with the decision to surgically reassign him as a girl being made at seventeen months. Perhaps, some reason, he was simply too old to make the transition, having been socialized as a boy for too long already. A second similar case, with the decision of sex reassignment occurring approximately between two and seven months of age, indeed has a more ambiguous outcome. The boy-who-became-a-girl has stayed that way, living socially as a woman.

During childhood, however, she recalls having been a "tomboy" and enjoying stereotypically masculine toys and games. She has had sexual experiences with both women and men, but is attracted predominantly to women in fantasy. She has a "blue-collar" job practiced almost exclusively by men (Bradley et al. 1998).

There are only two cases of this kind of accident (called "ablatio penis") that are reasonably well documented in the scientific literature. Two cases are not enough to allow us to draw any firm conclusions, especially since they point in partially different directions. There is another condition that might be considered even closer to the perfect nature-nurture experiment, however: cloacal exstrophy. Boys with cloacal exstrophy are born with poorly formed bowels that open into their bladder and with poorly formed penises. It is a very rare and very serious condition, and until the early 1970s babies suffering from it died shortly after birth. Now, due to the advancement of medical science, they can live. For some twenty years, most boys born with it were castrated and reassigned as girls immediately after birth. The development of fourteen of them has been followed up (last follow-up between ages fourteen and twenty), and these results definitely undermine the theory of psychosexual neutrality at birth. Seven of these female-assigned children have declared that they are boys. Five did that spontaneously and two after hearing from their parents what had happened. Of the seven others, one had wished to become a boy as a child but had accepted her status as a girl. Later, when her parents came clean about her birth, she became angry and withdrawn, refusing to discuss the matter. The parents of the other girls have decided never to tell their daughters. Follow-up reports suggest that these children are less than happy. All fourteen of these girls are sexually attracted to females, and all have unfeminine interests and behavior (Bailey 2003).

With regard to these and similar cases (for instance, cases of very feminine boys who state that they want to be girls), some environmentalists argue that maybe the child's gender socialization was inconsistent and ambivalent, due to the parents' knowledge of its birth status or due to their unconscious wish for a child of the other sex. Or maybe the psychologist who did the follow-up asked the children leading questions that caused them to question their gender identity. This would mean, however, that these very subtle features of interaction were much more consequential than calling a child a girl or boy and giving her or him a sex-typical name and sex-stereotyped toys and clothes. If gender identity development could really be undermined that easily, gender identity problems would be much more frequent (Bailey 2003).

Another phenomenon that is sometimes considered proof of the social construction of gender identity is the berdache. "[T]he Native American berdache undermines European notions of gender," Linda Nicholson

(1994:96) writes. What is a berdache? Nicholson provides this description: "[C]ertain Native American societies that have understood identity more in relation to spiritual forces than has been true of modern European-based societies have also allowed for some of those with male genitals to understand themselves and be understood by others as half man/half woman in ways that have not been possible within those European-based societies" (1994:96). Brouns notes that in some previous Native American cultures more than two sexes were acknowledged. "Certain individuals (male or female) to whom a privileged position was conferred could switch gender identity" (1995b:37, my translation). Sometimes they were men, then they were women again, Brouns says.

These accounts raise some questions. Both Nicholson and Brouns acknowledge that only *certain* individuals were allowed to understand themselves in that special way. Who were they? Why was not everyone allowed to do so? And could they just be half man/half woman or did they actually *switch* between a male and a female gender identity as they wished, which is something rather different? Did berdaches always have male genitals, as Nicholson seems to assert, or could male as well as female individuals be a berdache, as Brouns writes? And if only males could be so, then why were females not allowed to? Even without any further background information one thing seems for sure: Although in these societies gender identity may have been much more flexible than in the West, it was also subject to societal regulations, and only some people were allowed to be a berdache.

Can we infer from the existence of berdaches that gender identity is culturally constructed? This conclusion seems rather rash and does not seem to make sense. After all, berdaches have grown up in a society that ascribes different roles to males and females, just like societies everywhere do. If it were really that easy to make people digress from the sex-typical roles just by not reinforcing them into these roles, then that should work in our societies too. We know that it does not. It just is not that easy.

We need an expert here to supply us with the missing puzzle pieces. Historian Will Roscoe is. *The Zuni Man-Woman* (1991) is the account of Roscoe's nine-year research on alternative gender roles in Native American culture. By combining the methods of history and anthropology and by sharing his work with the Zuni, the culture he has been studying most intensively, he hopes to have avoided the pitfalls that troubled early anthropologist writings on this subject. Boasians like Ruth Benedict, in *Patterns of Culture* (1934), and Elsie Clews Parsons, a committed feminist, sought and found in Zuni culture a model to be copied. Having rejected the restraints that Western society imposed on genders, they had arrived with a predisposition to find Native American culture somehow more organic, Roscoe (1991) notes. It is to these ideologically predisposed anthropologists, seeking for unconventional ways of living, that we owe nearly all of our knowledge of Native American berdaches.

Although Roscoe's work is centered on the male berdache, he points out that there were female berdaches as well. A berdache was an individual who was allowed to act and dress as a member of the opposite sex. Female berdaches assumed male roles as warriors and chiefs or engaged in male work or occupations, whereas male berdaches engaged in domestic work. The phenomenon has been documented in over 130 North American Native American societies, in every region of the continent. Berdaches were integral and valued members of their communities, who had a very special and sacred place.

What makes the role of the berdache unusual to Western eyes is the high social status that was conferred to someone who would have been (or still is) considered deviant in our society. Still, according to Roscoe, it is clear that the individuals in case did not have flexible genders. They were lesbians or gays, whose inclinations were socially accepted, even sanctified. Male berdaches are not known to have had sexual relationships with women, but only with men. The key to defining berdaches, however, was less their sexual behavior than it was their preference for work typically done by the other sex. Berdaches were identified in childhood on the basis of their inclinations and skills. A boy who showed a preference for domestic work and association with women at the age of five or six was interpreted by adults as a berdache, so that at the age of ten he could wear a woman's dress. Thus people were not likely to label a child as a berdache without some demonstration of berdache inclinations.

Individuals, as Roscoe notes, are social actors, who do not passively await labels to trigger their behavior, but actively participate in shaping their identities and status. He firmly rejects the environmentalist interpretation of gender identity as a product of social influences and not as the manifestation of inherent drives. The existence of berdaches has been asserted as proof of the environmentalist paradigm, but this interpretation is at odds with the conclusions of Native Americans themselves. Some scholars even have gone so far as to deny the possibility of any connection between berdache and Western homosexual roles. Contemporary gay Native Americans, however, clearly see an overlap, but for them the berdache tradition encompasses spiritual and social dimensions as well as sexuality.

Gender identity, Roscoe (1991) concludes, is "constructed" in the sense that inherent sex and gender variant behavior is organized into formal social roles in different ways in different societies. To the extent that a child's actual interests corresponded with the conception of berdache status, this role provided a unique opportunity for personal growth, an opportunity that has historically been denied to Western homosexuals. I emphatically agree with Roscoe's argument that we can learn a lot by studying these cultural variations, since they can free us from single-dimensional definitions of homosexuality.

Thus the berdache tradition does not corroborate the environmentalist account of gender differences. It can be used, however, as a plea for the social acceptance of multiple kinds of gender identity, since it is clear that many people fall somewhere in between the two basic categories of male and female when it comes to gender identity. This observation, I may add again, does not detract from the evolutionary conception of our species as essentially dimorphic. The existence of variations on the basic theme is consistent with a Darwinian point of view; evolution is, after all, as much about variation as it is about the mainstreaming of the most successful phenotypes into the next generations. However, the continuum from anatomical female to anatomical male is not, as Fausto-Sterling (1992) seems to present it, a smooth uncurving line. In reality, most individuals are firmly located somewhere at one end of the continuum. As we saw, the number of children born with intersex genitals is not 4 percent, as Fausto-Sterling (1992) writes, nor 1.7 percent (Fausto-Sterling 2000b), but 0.018 percent (Sax 2002). The existence of these poles does, of course, not imply at all that society should try to force this rich anatomical and psychosexual variety into a binary sex and gender system, as we can learn from Native American cultures.

Environmentalist theories have no answer to the crucial question why children should learn so easily to behave in sex-typical ways, whereas their simultaneous socialization as prosocial and nonaggressive human beings turns out be much more difficult. It is precisely this question that is addressed by evolutionary psychology.

Notes

1. Allen 1997; Bleier 1985; Donovan 2000; Dusek 1998; Fausto-Sterling 1992, 1995, 1997, 2000a, 2000b; French 1992; Gray 1997; Haraway 1991b; Hubbard 1990; Hyde 1996; Nicholson 1994; Rose 1983, 1997, 2000; Rosser 1992; Sapiro 1985; Sunday and Tobach 1985; Talarico 1985; Tang-Martinez 1997.

2. Luckily not all left-wing intellectuals think that way. Noam Chomsky, for instance, thinks it important for political radicals to postulate a relatively fixed human nature in order to be able to struggle for a better society. We need a clear view of human needs in order to know what kind of society we want. He doubts, however, that science will be able to say much about it. According to him, we might rather try to find the answer to human nature in literature (Segerstråle 2000).

3. Yet tribal societies tend to have much higher homicide rates than industrialized nations, and in some European countries the homicide rate has declined spectacularly since medieval times. Today an Englishman's risk of being murdered is less than 5 percent of what it would have been in the thirteenth century (Daly and Wilson 1988).

4. Feminist biologist Lynda Birke implicitly acknowledges the factors laid out by Gross and Levitt when she writes: "Feminist science groups of the 1970s drew upon insights gained from other political issues; many of us involved, for example, were also engaged in the radical science movement, which had grown up in the late 1960s from the awareness of the part played by science in war and genocide, and global environmental damage" (1999:13).

5. For example, Browne 1998; Buss 1999; Daly and Wilson 1988; Gowaty 1995; Kimura 1999; Liesen 1995b; Lopreato and Crippen 1999; Low 2000; Nelissen 2000; Pinker 1997, 2002; Plotkin 1997; Pratto 1996; Ridley 1999, 2003; Roele 2000; Shermer 2001; Thornhill and Palmer 2000; Tooby and Cosmides 1992.

6. I realize that in this work I, too, have sometimes used "biological" in its more popular, dualistic sense. It just seemed a less cumbrous way of proceeding.

7. The controversy surrounding *A Natural History of Rape* epitomizes all that goes wrong in the relationship between feminism and the evolutionary sciences today. It is the perfect exemplification of all the problems touched upon in this book. For those readers who want to take a deep plunge in this heated debate: start by reading the book itself (of course). Let it be followed by *Evolution, Gender, and Rape* (ed. Cheryl Brown Travis, 2003a), a book volume that collects the reactions of a number of important critics. Then read Thornhill and Palmer's elaborate responses in which they painstakingly uncover the many misunderstandings and misrepresentations of their arguments (Thornhill and Palmer 2001; Palmer and Thornhill 2003a, 2003b—the first two can be found on the Internet). For a magnificent defense of an integration of life science and social science perspectives on rape causation, try getting hold of Owen Jones's article "Sex, Culture, and the Biology of Rape: Toward Explanation and Prevention" (1999).

8. The 1981 Penguin Books edition, *The Whisperings Within: Evolution and the Origin of Human Nature*, is an exact reprint of the original 1979 Harper & Row edition, but with a slightly different title.

9. This is suggested by Robert Kurzban (2002).

10. Abbey et al. 1996; Brouns 1995a; Brownmiller 1975; Bryson 1999; Butler 1990, 1997; Callen 1998; Connell 1995; French 1992; Hyde 1996; Nicholson 1994; Rose 1983; Rosser 1992; Segal 1999; ten Dam and Volman 1995.

11. The focus is here on the mammary glands, not on the prominent shape of women's breasts. In principle human females did not need to evolve permanent breasts in order to feed their babies. Nonhuman primates have enlarged breasts only when the female is lactating (Hrdy 1999b).

12. Sadly enough, David Reimer committted suicide in 2004.

5

Sociobiology and Evolutionary Psychology

> [A]pplied to females, pejorative-sounding words like "promiscuous" only make sense from the perspective of the males who had been attempting to control them—no doubt the origin of such famous dichotomies as that between "madonna" and "whore." From the perspective of the female, however, her behaviour is better understood as "assiduously maternal."
>
> Hrdy 1999b:87–88

DARWINISM HAS GONE A LONG WAY from the early sociobiological presupposition of essential differences between "ardent" males and "coy" females to the recognition that females are as dynamic strategists as males are and that female choice—which remains a cornerstone of sexual selection theory—is often constrained by male coercion. Although current Darwinian theory is not monolithic and although doubtless less-enlightened views of female nature will still be found,[1] I think that evolutionary theorizing on the sexes has become fairly balanced.

Anne Fausto-Sterling (1995) doubts whether the new, critical literature really dominates the field. It depends on what she means by that. If she means that female-oriented approaches are uncritically incorporated into the current body of knowledge, the answer is no. But if she means that most of these hypotheses are considered valuable points of view, potentially yielding new insights, and able to correct past biases, the answer is yes. Many evolutionary theorists are eager to investigate the fruitfulness of taking a female-centered approach, such as looking for mechanisms of mate evaluation in females, for instances of male circumvention of female choice, for mechanisms of female-female competition, and for the benefits of female promiscuity.[2]

In the end the ultimate value of these theories and hypotheses is decided by testing them—because most evolutionary hypotheses about behavior *are* testable, despite many allegations to the contrary.

This chapter deals with the development of and controversy about sociobiology and evolutionary psychology, a discipline that in the 1990s evolved out of sociobiology and the cognitive sciences. Although critics tend to dismiss evolutionary psychology as not being any different from its infamous predecessor (Dupré 2001; Hyde 1996; H. Rose 2000; S. Rose 2000), the difference is in fact quite big. As the name indicates, evolutionary psychology is in essence a science of the mind. It studies the role of the evolved architecture of the mind in mediating behavior. Sociobiology, in particular 1970s sociobiology, concentrates more on gene-driven behavior. It is an interplay of evolutionary ecology, population biology, and population genetics. Another difference is that evolutionary psychology does not assume human beings, nor any other organism, to be driven by the general goal of promoting gene survival. The evolved cognitive mechanisms that mediate behavior all serve some *specific* function and have been retained in the course of evolution because they just happened to help our ancestors to survive and reproduce (Symons 1992). Sociobiology, on the other hand, in the words of one of its early proponents, "proposes that our most important need is to maximize fitness" (Barash 1979:171–172).

Before starting to describe the historical developments leading to the rise of evolutionary psychology, one crucial conceptual distinction made within evolutionary biology has to be pointed out: that between proximate and ultimate explanation levels.

Proximate and Ultimate Levels of Explanation

Critics who do not understand the difference between the two levels of causation that are studied in evolutionary biology are, understandably, likely to infer that evolutionists intend an evolutionary approach of human behavior and an analysis in sociological terms to be mutually exclusive. Both approaches are, however, complementary.

Janet Hyde, for instance, writes that both biological-evolutionary and feminist explanations for the greater aggressiveness of males have been offered, but that feminist theories generally reject explanations that rest on biological factors. According to the feminist view, "the gender difference in aggression is a direct result of males being socialized into the male role and females being socialized into the female role" (Hyde 1996:113–114). She does not seem to recognize that *both* types of explanation may well hold, at a different level. Evolutionists do not deny the role of socializing, but they regard it as a direct, proximate cause that is in need of a broader level of explanation: an ultimate, evolutionary one.

Proximate causes refer to the immediate events that lead to a behavior or a certain trait. They include genes, hormones, physiological structures (such as brain mechanisms), and environmental (social, cultural, etc.) influences. Environmental influences are the ones typically studied by social scientists. The question always remains, however: *Why* do these proximate causes exist? We know that vision is enabled through, among others, a series of rods and cones in the eye, but why have eyes ever evolved? Behavioral differences between the sexes may be ascribed partly to hormones, but why do males and females have different hormonal levels in the first place? Socialization may be another part of the explanation, but why is it so easy to socialize children into the "appropriate" sex roles, whereas one will find it very hard to obtain the opposite purpose?

According to evolutionary psychologists, the answers to this kind of why-questions have to be sought in our evolutionary history. They are called ultimate or evolutionary explanations. *Proximate and ultimate explanations may, but need not conflict.* Both are needed if we want to completely understand any aspect of life.

Take Hyde (1996) again. She and Mary Beth Oliver conducted a meta-analysis of gender differences in sexuality.[3] They found a robust gender difference in attitudes about casual sex, with males typically looking for sexual gratification and females typically looking for emotional commitment. This is, indeed, what is to be expected from an evolutionary angle. Following Robert Trivers's parental investment theory, the less-investing sex is predicted to be less selective about mating than the more-investing sex. Hyde writes that "[s]ociobiology, social learning theory, social role theory, and script theory all predict this gender difference" (1996:115). In the rest of her argument she seems to assume that either it is a matter of biology or of gender-role socialization. Since gender role socialization, and the sexual double standard in particular, have both predicted and can explain the phenomenon, we do not need an evolutionary approach, she reasons.

I believe she is wrong. This is like saying that languages vary between countries and ethnicities because people are raised in different languages. Indeed, that is one level of the explanation, and it explains why a French-speaking community does not turn into a Japanese-speaking one overnight. But clearly we do not get a firm grasp of the phenomenon of differing languages if our analysis remains stuck at this level. How have languages ever come to develop in different ways? We need a historical approach here. And how has speech itself ever come to evolve? Only an evolutionary approach can provide an answer to this. We will have to clarify not only why our species has evolved the ability to speak, but also how the intricate process of attuning this mechanism to the cultural specifics works. All these levels of analysis are complementary. Lynne Segal is fighting a nonexisting enemy when she writes that "[t]here is not, and never could be, any single, unified project with the capacity to encompass the different levels of explanation necessary for understanding the complexity of human affairs" (1999:113). No one is suggesting that the complexity of human life could be reduced to one level of analysis.

Socialization theory can only predict how gender roles will affect people if one already knows what these roles are. And yes, socialization matters a lot—something no evolutionary theorist will deny. What socialization theories cannot explain, however, is how these gender roles came to be in the first place, and why the same basic pattern of sex differences is found in all known cultures—although the cultural expression of them can vary tremendously. The theory of evolution by natural and sexual selection provides a broader framework that does not only explain the ubiquitousness of these sex differences in many species, but can also predict them—although, as I have stressed quite a few times, it took the contribution of (mainly) women researchers to point out the ways in which female behavior had often been assessed from a patriarchal point of view.

An ultimate account of behavior (which posits that the behavior exists because it was favored by selection) does not at all deny that learning is involved in its occurrence (which is a proximate account). Yet there is sometimes an incompatibility between our knowledge of how evolution works and the specific proximate causes proposed by social scientists. The Freudian Oedipus complex, for instance, could never have evolved as a fundamental part of human nature. Because of the reduced viability of offspring produced by incestuous mating, the trait would soon have died out. The Oedipus complex, one of the core presumptions of psychoanalytic theory, would never have been given any credence if anyone had considered the evolutionary fate of a trait that produced such incestuous desires in each and every child. The same goes for Freud's proposition of the death instinct, and for many other psychoanalytic concepts as well.

The process of natural selection thus constrains the number of traits that could possibly exist as evolved psychological adaptations. This observation can spare us a lot of fruitless theorizing, such as in the debate on the origins of sex differences.

To ask *why* a given trait or behavior exists is a legitimate scientific question. It need not be proof of teleological thinking. Darwinian theory is only *seem-*

ingly teleological, because the consequences of biological phenomena constitute an essential part of their explanation. If wings had not conferred evolutionary advantages, they would have been extinguished at an early stage of their development. Why an organ or a preference or a behavior exists is just as legitimate a scientific question as how it works. Answers of the form "in order to . . ." are legitimate scientific answers. We can say that the liver exists for detoxification, our preference for sweet foods for promoting the ingestion of nutritive sugars, and male sexual jealousy for paternity insurance (see next chapter) (Daly and Wilson 1996). Natural selection is a blind mechanism, but this does not mean that traits are retained in a haphazard way: They are only retained if on average they promote or do not compromise the survival or reproduction of the organism in one way or another.

A common confusion, however, is to imagine that fitness *itself* is what people and other animals strive for. An evolutionary explanation does, however, not imply a conscious intent to reproduce. As the offspring of successful forebears, individuals are predicted to be *adaptation executioners* rather than maximizers of personal reproductive success (Tooby and Cosmides 1992). A distinction has again to be made between explanations at the proximate and at the ultimate level of behavior. An example: People like to have sex because sex is fun. This is a very good proximate answer to the question: Why do people have sex? The pleasure we experience while having sex, however, evolved "in order to" ensure that we would do what is necessary in order to propagate our genes. This is an ultimate answer to the question why people have sex. Both answers are right; they are just complementary.

With our modern methods of birth control, sex and reproduction are no longer automatically linked. Sex, however, is still fun. Our bodies and minds have evolved over millions of years in an environment without birth control and they have retained the traits that were adaptive then, although these may no longer be adaptive now. Because evolution occurs slowly, existing humans are necessarily designed for the previous environments of which they are the product. Thus the currently nonadaptive or actually maladaptive character of some traits (such as our desire for fat and sugar) does not imply that they cannot be adaptations.

Sociobiology: Of Genes and Men

The official birth of sociobiology is dated at 1975 with the publication of Edward Wilson's ambitious and notorious book *Sociobiology: The New Synthesis.* In it Wilson had drawn together theoretical ideas (such as those developed by William Hamilton and Robert Trivers) and empirical studies (from the fields of ethology, population genetics, and behavioral genetics) from the last decades in one huge construction, by which animal social behavior was viewed as genetically controlled and evolving through natural selection. Although Wilson did not say anything fundamentally new, his important contribution was that he gave the emerging field a name and that he advocated its importance in a social climate suspicious of evolution and the genetics of behavior (Segerstråle 2000).

When seen against the background of the environmentalist paradigm, which was then (and still is) predominant and considered politically correct, it is not surprising that the book caused a lot of controversy. Wilson owed the uproar partly to himself, however. Of the nearly 700 pages, the last 30 were devoted to humans. Empirical evidence on the genetic underpinnings of human behavior was scarce in those days. Still he asserted that sociobiology would "cannibalize psychology" (1975:575) and suggested that the other social sciences would suffer the same fate.

Wilson made other dubious claims, such as putting forward the possibility that the time had come "for ethics to be removed temporarily from the hands of the philosophers and biologicized." One has, he thought, to consider "the ultimate ecological and genetic consequences" of ethical systems (1975:562). Whereas these are vague assertions—Wilson's position on ethics remained rather unclear and self-contradictory throughout the years (Segerstråle 2000), they seem to come dangerously near the naturalistic fallacy.

In this respect Wilson differs tremendously from Dawkins (1976), who explicitly disconnects values from facts, as most sociobiologists do. Segerstråle (1992) notes that feminist critics of sociobiology in general have, however, not cared about the considerable differences in content and approach between sociobiologists. They seem to use 'sociobiology' as an umbrella term for practically any study of sex differences, be it psychological, physiological, or biological. Their criticisms have, moreover, remained remarkably similar during the last decade, although evolutionary theorizing has become much more sophisticated.

Since early sociobiology was an integration of ecology, ethology, and evolutionary biology, it is not that strange that it inherited some of the androcentric points of view that at that time characterized those disciplines. They have already been addressed throughout this work: the focus on dominance relationships and on the importance of males in organizing social life; the stereotyping of females as relatively monogamous, passive, and coy; the neglect of female interests and strategies; the supposed pivotal role of male-male competition in the evolution of a species; and, with respect to humans only, the focus being more on the evolutionary importance of men as hunters than on women as gatherers.

Although female choice was gradually coming to be more accepted in the 1970s, it would still take some time before female interests and strategies were conferred the same status as those of males in evolutionary accounts of behavior. As we saw, various societal and theoretical developments came together to produce this state of affairs. Feminism led to more women studying and contributing to biology, it probably informed the theoretical perspectives of some of them, and it created a society in which female choice did not seem so strange anymore. There was Robert Trivers's 1972 seminal article on parental investment, which emphatically confirmed the evolutionary importance of female choice. There was the field work of women primatologists—some of them feminists—attesting to the correctness of their suspicion that a lot of earlier work in primatology had been too much a reflection of a patriarchal society. There were also the (feminist and other) critics outside the field who, according to Segerstråle (2000), unintentionally helped strengthen the sociobiological approach, since because of their criticisms, claims about sex differences and about other issues had to be increasingly carefully substantiated.

Sarah Hrdy and the Lusty Female

Hrdy's 1981 *The Woman that Never Evolved* (1999b) was of critical importance in challenging several assumptions about female mating behavior. Sociobiological theory maintained that females would tend to be relatively monogamous since, having a quite limited capacity to conceive, reproductively they could win nothing by mating around. Males on the other hand always had the possibility of doing better. Since a male's reproductive success was thought to be dependent upon his eagerness to copulate with any fertile female, selection would have favored sexually assertive males. But, so the argument ran, since copulation serves no function other than insemination and since females in a natural state breed at or near their reproductive capacity, selection would not favor the evolution of a sexually assertive female (Hrdy 1999a).

Hrdy pointed out, however, that actually the females of many primate species are lusty creatures. They solicit sex, actively select their mates, and are willing to take risks in order to mate with many more males than is necessary for insemination. A chimpanzee female, who will have no more than five offspring in a lifetime, will solicit thousands of copulations from dozens of males. Many female primates also engage in sex while not ovulating, which means the chance of conception is minimal. The undeniable eagerness of females in a wide array of primates to engage in nonreproductive copulations with a lot of partners needs an explanation, Hrdy asserted.

Until then, it was supposed that all nonhuman primate species showed strictly cyclical sexual behavior (the period of sexual activity is known as estrus). The continuous receptivity of human females and other womanly attributes such as prominent breasts and buttocks were most commonly explained as a service to males. With Desmond Morris as the most prominent spokesman for this view in his 1967 book *The Naked Ape*, the human female was assumed to have become uniquely sexualized in order to attract and keep her man, so that he would stay with her and help her rear the highly dependent babies (Hrdy 1999a). Although Hrdy assumes that there is some connection between pair-bonding, paternal investment, and the evolution of continuous female receptivity, her evidence about the sexual behavior of other primates showed that that is not all there is to it, and that the prevailing view of female primate sexuality, human as well as nonhuman, had been extremely male focused.

She showed the sexual sophistication of the human female to be at the highly erotic end of a primate continuum, thus not necessarily demanding a special explanation applicable *only* to humans (another manifestation of this continuum is the observation that an increase in libido occurs among women at midcycle, just as it does in other primates). She pointed out that, instead of females in a natural state breeding at reproductive capacity, there is in fact considerable variance in their reproductive success. This leaves room for natural selection to operate on them. Female strategies that lead more offspring to survive will in the long run become widespread among the population. Female promiscuity may under certain circumstances be such a strategy: Through mating with more than one male, she makes paternity uncertain for the male. This may be a way of preventing infanticide by males and of extracting additional resources for her offspring (Hrdy 1999a).

While to some this may sound like an untestable proposition, it is not. In Hrdy (1999b) we find the results of some research on paternity in langurs and dunnocks. The langurs of Nepal live in harems with frequent overtakes by new males, with most males living as nomads and awaiting their chance to take over a harem. Infanticide by new troop leaders accounts for 30 to 60 percent of all infant mortality. DNA evidence indicates that *none* of the infants killed could have been sired by the males who killed them. The mother's past relationship with the male (before he became troop leader) provides the cue for him to either tolerate or attack a particular infant. Thus by mating with nomad males, who might become troop leaders later on, a female langur seems to try to protect her future offspring from infanticide, since a male virtually never attacks an infant born to a female with whom he has mated, even if she has mated with other males as well.

Evidence that male animals are capable of remembering whether they mated with a particular female comes from European sparrows called dunnocks. Female dunnocks live in cooperative breeding groups in which a female solicits multiple males. These males, in turn, help provision the chicks more or less in proportion to how much opportunity they had to inseminate the female when she was last fertile. Males are significantly more likely to bring food to young they fathered or young they might have fathered. DNA evidence revealed that males were often *but not always* accurate in assessing their fatherhood.

A female, Hrdy (1999b) concludes, is doing much more than just selecting the one best male from available suitors. She is actively manipulating information available to males about paternity, in order to protect her offspring and to get resources. By being promiscuous, she is doing all she can to secure the survival of her young.

In the next chapter I take a further look at current theories about the benefits of female promiscuity, as well as at the existing evidence for them.

Other Feminist Critiques of Sociobiology

The list of feminist critiques of sociobiology is long. The discipline was (and is) charged with being fundamentalist, sexist, genetic determinist, unscientific, reductionist, making selective use of animal samples, falsely universalizing behavior, making political statements, and using sexist language. Feminists also objected to the whole notion of animal analogies to human behavior.[4]

Some of these critiques were warranted at the time. There was little hard evidence on humans and analogies to animal behavior were sometimes made without caution. Today it is recognized that one should not look at isolated cases but try to draw principles from a whole range of species, try to discern the patterns in it. Under what circumstances, for instance, do males defer to females? (Hinde 1987; Hrdy 1999a). Another example of a—theoretically predicted *and* found—pattern is that across animal species paternal care will be small or nonexistent when paternity certainty is low, when it does not substantially improve offspring survival rates, or when it severely compromises mating opportunities with other females (Geary 2000). Also, across species with internal fertilization the less-investing sex—typically the males—is predicted and found to try to regulate the female's copulation behavior in order to ensure paternity (Trivers 1972; Birkhead 2000).

A particularly telling example of a pattern is that of the relationship between relative testes size and level of promiscuity in primates. A neat correlation can be found between sperm volume as indicated by testicular weight corrected for body size and degree of sperm competition (that is, the likelihood that the female has recently mated with another male, with her reproductive tract still containing competing sperm). The function of producing a large amount of sperm is hypothesized to be that of displacing rival sperm, hence increasing the odds of fertilizing a female's egg. Gorillas, who guard their harem of females very intensively and thus do not have to fear rivals, have very small testes (accounting for 0.02 percent of body weight). They are followed by orangutans (0.05 percent of body weight), with orangutan females being slightly promiscuous. Chimpanzees, who are not able to keep females from mating around, have very large testes (0.27 percent of body weight). Human testes are larger than those of orangutans, but much smaller than those of chimpanzees (0.08 percent of body weight), indicating that our female ancestors were not monogamous. They were more promiscuous than orangutans, but not nearly as promiscuous as female chimpanzees (Hrdy 1999b; Miller 2000b).

Segerstråle (1992) notes that the critiques in fact address fundamental problems of science in general and biology in particular, while formulated as attacks on individual scientists. She thinks that the sociobiology controversy was not only a nature-nurture conflict, but that it was also a matter of clashing conceptions of good science, particularly with regard to the study of human behavior. To sociobiologists there was nothing wrong with being speculative, nor with deliberate oversimplification. They believed this to be crucial for theory formation. In their eyes the criterion for a good theory was testability and hence they felt entitled to draw conclusions about sex differences in behavior on the basis of physiological or psychological research.

Some critics on the other hand believed such studies to be inappropriate for our species. When studying humans one must act in a purist way: not working with assumptions, but only with data that are known to be true. What sociobiologists considered creative science, the critics considered bad science. To them it seemed clear that sociobiologists could only be choosing to do "bad" science out of political motives, whereas sociobiologists believed that errors would be filtered out simply by doing *more* science (Segerstråle 1992, 2000). And this is indeed what has gradually been happening during the past few decades.

All of the earlier-mentioned charges are repeated today with respect to evolutionary psychology and current sociobiology, in spite of current theory being much more sophisticated, evidence based, and female oriented. I focus on one typical habit of the critics: what Segerstråle calls "massaging texts through moral reading" (2000:208).

Moral Reading

According to some feminist critics (Blackman 1985; Bleier 1985; Fausto-Sterling 1992; French 1992; Hubbard 1988, 1990; Segal 1999), an evolutionary account of sex differences may well be *presented* as a mere explanation of the existence of sex roles, but what sociobiologists *really intend to demonstrate* is that it is impossible to establish a society where true sexual equality exists. What is remarkable is the selective way in which the critics scrutinize the target texts in order to arrive at their preferred interpretation.

According to Segerstråle (2000), the basic idea behind this kind of 'moral reading' is to imagine the worst possible political consequences of a scientific claim. In this way, maximum moral guilt might be attributed to the perpetrator of this claim. Because the critics "know" what evolutionary theories are necessarily about, they "help" the reader by teasing out the "true" underlying messages. This guided reading starts with the necessary assumption that evolutionary theorists have a political point to prove, after which all nonpertinent quotations are ignored, and only the ones that bring out the true meaning are presented out of context.

Take the way in which some of Wilson's quotes on sex differences are taken as proving his "patriarchal agenda" (French 1992:124). Often Wilson's position is represented in a misleading way, and sometimes in a downright false way. We find Marilyn French saying that Wilson "asserts 'males are dominant over females'" (1992:122), whereas he actually states that "males are *usually* dominant over females" (1975:291, emphasis added) and thereby offers examples of species in which females are dominant, such as the hyena, the vervet monkey, and Sykes' monkey. With regard to Wilson's treatment of the territorial behavior of pike blennies and his observation that female blennies are not challenged by the males, she writes that he "interprets this to mean males control females" (1992:123). In fact the only thing Wilson writes is that "[v]ery probably the tolerance toward them is the prelude to courtship during the breeding season" (1975:260). Nowhere do we find Wilson writing that he thinks males are "superior," as French asserts he does (1992:123). Obviously French tailors her interpretation to fit her preconceived scheme of things.

Only with regard to Wilson's choice of language do her critiques seem reasonable. The use of the word "harem," for instance, may often be misapplied when discussing all-female societies where females associate with males only at mating season or societies where females live together with one resident male. It is possible that females created such an arrangement to avoid male aggression (Smuts 1995, 1996). Wilson's description of female sexual posture as "receptive" or "submissive" is another example of word use that is rightly criticized by French, because it seems to imply that penetration equals male domination.

To infer on this small linguistic basis that Wilson is defending a patriarchal agenda does, however, not seem to be justified. A more moderate, sensible interpretation would be that Wilson's word choice unwittingly reflects the cultural climate of that time. That interpretation gets reinforced if one knows that Wilson actually considers himself a feminist (Segerstråle 2000). Indeed, one obvious way to assess the correctness of the critics' assumptions about their targets' political motivation would be to see whether these hold up *empirically.* But as we have seen, the connection between right-wing political interests and sociobiology or evolutionary psychology, however politically useful it may be for the critics, does not hold up in practice.

Wilson's elaboration on sex differences in his 1978 book *On Human Nature* has been another object of moral reading. He suggested that "at birth the twig is already bent a little bit" (1978:132) and that, as a society, we can choose what to do with that. He offers three possible ways to follow: We can condition people so as to exaggerate sex differences in behavior, as almost all societies do; we can train them so as to eliminate these differences; or we can just provide equal opportunities but take no further action. All three options, he points out, have their social benefits *and* their costs. The only firm conclusion to be drawn, according to him, is that "the evidences of biological constraint alone cannot prescribe an ideal course of action," but "they can help us to define the options and to assess the price of each" (1978:134).

Wilson clearly writes that it is "the modest *predisposition* toward sex role differences" that is "unavoidable" (1978:135, emphasis added). Fausto-Sterling (1992) takes this as evidence that he is defending the status quo. In a later interpretation this becomes: "He sets a price on malleability's head and tells us that this is too dear for most of us in a democratic society to be willing to pay" (Fausto-Sterling 2000a:184).

No such thing is asserted by Wilson. He is just conducting an analysis of different alternatives and their consequences. One would think it is a wise strategy to assess the options in a relatively dispassionate manner before starting to reform humanity by trying to erase all sex differences. We might want to try and learn something from other large-scale historical attempts at reforming human nature, such as those in Stalinist Russia and in China under the Cultural Revolution, which turned out to be a nightmare. The ambition to remake human nature turned Stalin and Mao into dictators and mass murderers. There have been attempts to create a new and better society with less terrible results—utopian communities, the Israeli kibbutzim—but none that are unqualified successes. The communes that sprang up in the United States and Europe in the nineteenth century and early decades of the twentieth all collapsed from internal tensions. The collectivist philosophy of the kibbutzim was undermined by their members' desire to live with their families, to own their own clothing, and to keep things acquired outside the kibbutz (Pinker 2002).

From an evolutionary point of view, human behavioral flexibility is expected not to be boundless. That does not mean that flexibility is not considered a crucial asset of our species, and evolutionary psychology stresses it still more than does sociobiology.

Cognitive Science Setting the Stage for Evolutionary Psychology

In the 1960s a deep change took place in psychology. Radical behaviorism was refuted by many experiments showing that the external environment is not the sole determinant of behavior and that organisms come equipped with particular types of learning. Learning language, for instance, comes easily in children, although the rules are very complicated. This ability stands in sharp contrast with learning how to read or carry out divisions. We obviously learn a lot of complex things with seeming ease: making coherent sense of a mass of information coming through the eyes, for instance, or performing everyday but highly complicated motor maneuvers (Buss 1999; Pinker and Bloom 1992; Plotkin 1997).

Behaviorism gave way to cognitivism, which became the dominant approach in psychology and has remained so ever since. Cognitivism accepted the existence of hidden causes of behavior and made it acceptable again to have recourse to them in psychological theory. These hidden causes were, and are, assumed to be identifiable with brain structures and functions. Noam Chomsky, for instance, argued that the spontaneity, rapidity, and uniformity of language learning in children can only be explained by the existence of neurological structures that are the source of this rule-governed process. Similarly it was argued that certain features of human memory can only be explained by the presence of some information-processing device. Both kinds of neurological device are examples of structures of the mind that in nontrivial ways cause our behavior to be what it is (Plotkin 1997).

In cognitive science (and in evolutionary psychology) the word "mind" refers to an information-processing description of the functioning of an organism's brain. With "information" any psychological process can be meant, such as reasoning, emotion, motivation, or motor control. Cognitivism is about information processing: It describes what kinds of information an organism takes as input, what procedures it uses to transform that information, what kinds of data structures (representations) those procedures operate on, and what kinds of representations or behaviors it generates as output (Tooby and Cosmides 1992).

What the early cognitivists did not do, however, was ask about the origins of mental devices such as the neural structures underlying learning and intelligence. Moreover, most cognitivists carried over the behaviorist assumption that the mind is domain-general. Human cognitive mechanisms were seen as constituting a large computer designed to process any information it was fed. The idea that the human mind might be biased to attend to specific kinds of information did not occur. Whereas in behaviorism the mind was regarded as a blank slate or a sponge, it was now seen as a general-purpose computer (Buss 1999).

The mind, however, evidently is not indiscriminate in the knowledge it accumulates, nor does it simply solve problems in the way a computer does. Just contrast the ease with which children acquire language with the effort it takes them to learn how to read, write, and solve abstract mathematical problems, and it will become clear that the human mind is not a general-purpose computer. It is a creative system that seems to contain lots of specialized devices for solving specific and highly complex problems, such as making sense of speech sounds, of the visual environment, and of the intentions of other human beings. In the 1980s many cognitive psychologists started to address the nature and number of these devices, which were called "modules." Jerry Fodor was the first to introduce this idea of modularity in his 1983 *The Modularity of Mind.* The existence of modules was attested by numerous lines of evidence, including their linkage to specific brain regions, their characteristic patterns of development in children, and the patterns found when they break down. Face recognition, for instance, seems to be a separate module within the visual system. This is strongly suggested by the fact that newborn infants automatically attend to facelike patterns; that humans are capable of recognizing a very large number of faces while being very bad at describing them with words; and that prosopagnosia, the inability to recognize faces, arises from damage to a specific region of the prefrontal cortex (Mithen 1996; Plotkin 1997).

The Mind as a Swiss Army Knife

The idea of modularity was developed further by evolutionary psychology, which is simply psychology informed by evolutionary biology. It argues that we can only understand the nature of the human mind by viewing it as a product of the evolutionary process. The mind is such a complex, functional structure that it could not have arisen by chance. The only known process that can organize complex functional design, that is, complex design that can solve an adaptive problem, is evolution by natural and sexual selection. Evolutionary psychology thus wants to supply the necessary connection between evolutionary biology and the complex, *irreducible* social and cultural phenomena studied by social scientists.

Evolutionary psychologists assume that the information-processing mechanisms that constitute the human mind are adaptations: They evolved as a way of coping with recurring problems of survival and reproduction, such as selecting the right foods to eat and the right mate with whom to have children, in the course of the two million years our ancestors spent as Pleistocene hunter-gatherers. This world in which our ancestors evolved is called the environment of evolutionary adaptedness or EEA and must be seen as a statistical set of selection pressures rather than as an actual physical environment. Ecologies have changed a lot during that long time period. Some aspects remained stable, however—the laws of gravity did not change—and some problems recurred with sufficient regularity to produce specific problem-solving devices, such as the problem of group living and social cooperation. It is unlikely that new *complex* designs could have evolved in the short time span since the appearance of agriculture 10,000 years ago—equaling 400 human generations—since natural selection does not work that fast (Cosmides, Tooby, and Barkow 1992). If they had evolved, the psychology of extant human-gatherer populations would be significantly different from that of inhabitants of industrialized societies. It is not (Buss 1999).

The view of the mind advanced from an evolutionary framework is that of a Swiss army knife, consisting of hundreds or thousands of evolved psychological mechanisms or mental modules. Just as our bodies contain thousands of specialized mechanisms, directed at the solution of specific adaptive problems (we have lungs for oxygen uptake, a liver to filter out toxins, a heart to pump blood, etc.), so does our mind—which is, after all, a physical part of our body (Humphrey 1992, 2000). It is this multiplicity of domain-specific mental modules, guiding us to correct adaptive solutions to adaptive problems, that gives rise to the tremendous flexibility of behavior that characterizes modern humans (Buss 1999).

Although it may seem sensible to think that a general problem-solving mechanism is more flexible than are content-specialized mechanisms, since the organism is not kept from doing what is desirable (or adaptive) by a "rigid" architecture, this view is wrong. David Buss (1999) suggests that we think of a carpenter's toolbox. The carpenter gains flexibility not by making use of one highly general tool, but by having many specific tools in his toolbox. These can then be used in many combinations that would not be possible with the one general tool.

Indeed, cognitive and artificial intelligence researchers consistently find domain-general systems too weak to solve virtually any complex real-world task. The reasons for this can be expressed by two related concepts, called by artificial intelligence researchers and other cognitive scientists 'combinatorial explosion' and the 'frame problem'.

Combinatorial explosion refers to the fact that with each new degree of freedom added to a computational system, the total number of alternative possibilities that it faces grows exponentially. The system cannot possibly compute the anticipated outcome of each alternative—something that is called the frame problem—and thus must make a choice without complete consideration of the overwhelming range of possibilities. The frame problem is known by different names in linguistics, semantics, developmental psychology, and perception as well. A system that only has a domain-general architecture could never solve this problem. To be sure, it *does* need general problem-solving techniques, such as the ability to make associations, the ability to recalibrate based on the consequences of actions, and the ability to reject propositions because they are contradicted. But it also needs specialized problem-solving techniques that are only activated when they encounter the domains to which they are applicable, such as faces, grammar, or falling in love. Any past design for an organism that could not generate appropriate decisions or perceptions would not have propagated and consequently would have been removed from the population in the next generation.

The frame problem was illuminated for the first time by artificial intelligence researchers. To their great surprise, they found that it was very difficult to discover methods that would solve problems that are easy to humans, such as seeing or moving objects. To get their programs to handle even absurdly simplified tasks such as moving a few blocks around, they had to build in substantial "innate knowledge" of the world. To move an object, make the simplest induction, or solve a straightforward problem, the computer needed a sophisticated model of the domain in question. It needed content-rich and specialized frames. A frame provides a worldview; it carves the world into defined categories of entities and properties, defines how these categories are related to each other, provides methods for interpreting observations, suggests what information is lacking and how to get it, and so on. Whether talking about an artificial intelligence system or an evolved organism, the problem is the same: The world cannot supply to the system what the system needs first in order to learn about the world. Therefore the essential kernels of content-specific frames must be supplied initially by the architecture. In humans these content-specific mechanisms (modules) are, among many others, a face recognition module, a spatial relations module, a fear module, a social-exchange module, an emotion-perception module, a kin-oriented motivation module, an effort allocation and recalibration module, a child-care module, a sexual-attraction module, a semantic-inference module, a friendship module, and a grammar acquisition module. It is difficult for us to imagine, however, that these frames are indispensable prerequisites for functioning as a human being, precisely because they make us feel as if we steer through the world effortlessly (Tooby and Cosmides 1992).

Hence behavioral flexibility does not mean that responses are unlimited. It means the capacity to adjust behavior in an appropriate or successful way to the specific circumstances faced. Thus evolved structure does not constrain; it creates or enables. It is because of, and not despite, this structure that our behavior is so context-sensitive. Some social scientists, however, think that both notions of flexibility (flexibility as the absence of any limits on responses and flexibility as contextually appropriate behavior) are the same, or that the first kind automatically leads to the second kind. The problem of combinatorial explosion and the frame problem show that this cannot be true (Tooby and Cosmides 1992).

It is not, however, because life's machinery is designed by natural selection that it incarnates some generalized reproductive striving. People want many things, but they do not want such a hazy thing as "to reproduce." They long for sex because they have a sex drive and because sex feels nice. They may long for children because the idea of having a child appeals to them. But that is not the same as wanting a child because of the desire to *reproduce* (this, by the way, makes it possible that people adopt children and love them as their own).

One must not confuse the general process that produces adaptations with the adaptations themselves. No mechanism can serve the general function of promoting gene survival because there is no general, universally effective way of doing so. Adaptations serve *specific* functions, such as regulating blood pressure, perceiving edges, detecting cheaters in social exchange, and thousands of other things. Selection acts in such a way that the things that enhance our survival or reproduction tend to be experienced as pleasurable. Thus human beings, like all other organisms, have been designed by selection to strive for specific goals, such as keeping warm, finding a good mate, achieving social status, and having sex. Our ways to achieve these ends are, however, uniquely flexible, and the psychological mechanisms underpinning this flexibility make possible new means to the same ends. Our love of sweet foods, for instance, was shaped by natural selection in prehistoric times, when high-calorie food was scarce. Flexible as we are, we have been able to develop many ways of obtaining and concentrating sugar, but the goal of eating it remains the same: to experience the sensation of sweetness (Symons 1992).

The evolved architecture of the brain processes environmental information in a way that guides feelings and behavior toward ends that were adaptive in past human environments, but that does not mean they are still adaptive today. Take again the example of sugar. In ancestral populations, sweet foods were so scarce that problems of obesity were improbable to arise. In industrialized societies today we no longer need a sweet tooth in order to survive, because there is food in abundance. But we have retained this preference for sweet tastes, with fatness and teeth filled with cavities as a result (Low 2000).

Indications of a Panhuman Design

No single discovery can prove that there is a universal, panhuman design of the mind stemming from our hunter-gatherer existence, but many lines of evidence converge on it. Findings from anthropology, evolutionary biology, cognitive science, linguistics, comparative psychology, neuroscience, developmental psychology, and artificial intelligence research all point in the same direction: that of an evolved and specialized architecture of the mind—be it that men and women have evolved to behave differently on a psychosexual level (Brown 1991; Buss 1999; Geary 1998; Low 2000; Pinker 1997, 2002; Tooby and Cosmides 1992). We have a look at a few examples stemming from anthropology, developmental psychology, and neuroscience.

Human life and human societies are full of structural patterns all over the world. In *Human Universals* (1991) anthropologist Donald Brown describes more than 400 statistical regularities recurring in every known culture. Humans have characteristic emotions. They live in groups and often favor their close kin. They have a sense of being a distinct people, and important conflicts are structured around in-group-out-group antagonisms. They distinguish right from wrong. Their ideas include a worldview. They compete over limited social or subsistence resources. They create enduring, mutually beneficial relationships with nonrelatives. They use language to organize, respond to, and manipulate the behavior of others. Their speech is highly symbolic. They gossip. Their kin terminology includes terms that distinguish male from female and some generations from others. They have a sex terminology that is fundamentally dualistic, even when it comprises more than two categories; in that case the third (and sometimes fourth) category is a combination of the two basic sexes or a crossover sex. They know romantic love, sexual attraction, and sexual jealousy. Sex is a topic of great interest to them, and there are societal regulations with regard to it. They have a concept of the person in the psychological sense. They know that people have a private inner life. They are tool makers and use tools to make tools. They have rituals, including rites of passage. They have standards of sexual attractiveness, including signs of good health and a male preference for female youth. They have leaders, although these may be situational or ephemeral. They have statuses and roles. They show affection as well as feel it. They have laws proscribing violence, murder, and rape (although they may justify these actions in some contexts). They have a division of labor, minimally based on sex and age. Women have more child-care duties. Men are dominant in the public political sphere and are more physically aggressive. The list goes on and on. Tooby and Cosmides (1992) conclude that our immensely elaborate species-typical physiological and psychological designs not only constitute regularities in themselves, but that they also impose within and across cultures all kinds of regularities on human life, as do the common features of the environments we inhabit.

The knowledge that evolved cognitive modules generate is always *implicit*: It is spontaneously inferred by regularities in the outside world, without awareness of the underlying principles. Young children everywhere intuitively recognize that living things and inanimate objects are fundamentally different, for instance. They automatically attribute an "essence" to different kinds of living creatures; thus they will still recognize a cat with only three legs as a cat. In all societies children seem to have this kind of intuitive knowledge about the world in at least four domains: language, psychology, physics, and biology. Their social behavior indicates an implicit understanding of the costs and benefits of reciprocal relationships. They do not have to be taught Euclidean geometry in order to know that the shortest distance between two objects is represented by a straight line, or to be able to navigate through a room. Infants everywhere spontaneously cut up the strings of sounds that come from their parents' mouths into intelligible units. They have a rudimentary understanding of the emotions signaled by facial expressions. From a very early age they grasp the properties of physical objects: solidity, gravity, and inertia. They understand physical causality. No explicit instruction is needed for these cognitive abilities to develop, and they appear too early in development to result from the infants' experience of the world. Not surprisingly, they all concern relationships that reflect the structure of the world as it has been for thousands of ages or longer. In other words, it concerns knowledge that is directly related to a hunter-gatherer lifestyle. The classification of plants and animals, for instance, is similar in all cultures and turns out to be highly consistent with the scientific categorization of these species (Geary 1998; Tooby and Cosmides 1992).

This implicit understanding of the structural properties of the physical, biological, and social world appears to be represented in the structure and functioning of the underlying neural systems. In the view of many developmental researchers, most of these innate competencies are only skeletal at birth and are refined through the child's interactions with its environment. "[I]nnate skeletal knowledge biases the types of information that infants and children attend to, influences the ways in which this information is processed and represented, and motivates infants and children to engage and to learn about their local ecologies." Also, "an inherent motivation to seek out environments that correspond to the evolved modules . . . is expected" (Geary 1998:200). Although the specific architecture of these cognitive modules is not completely known, it seems clear that most of them are open systems, designed to be influenced by contextual factors, especially during childhood (Geary 1998).

Another example of an evolved cognitive module is the 'theory of mind' module. At roughly three years of age children across cultures develop a theory of mind: the capacity to make inferences about the beliefs and desires of other people. Cognitive psychologist Nicholas Humphrey (1976) has proposed that individuals with an ability to "read" other people's minds would have achieved a relatively greater reproductive success, since they would have been able to predict the behavior of others. Thus we can identify selection pressures for a theory of mind module and we find evidence for its existence in developmental psychology. Moreover, evidence from cognitive neuroscience suggests its localization in the brain. Developmental psychopathologist Simon Baron-Cohen (1995) showed the condition of autism, in which individuals seem to be unaware of what other people are thinking and feeling, to probably arise from an impairment of this module. Having no theory of mind means not being able to empathize, to put oneself into someone else's shoes. In an extension of this approach, Baron-Cohen (2003) considers mind-reading in the broader context of sex differences in empathy. He describes the autistic brain as "the extreme male brain": Autistic individuals have impaired empathizing, a quality in which females typically show an advantage, but are obsessed by systemizing, a quality in which males typically excel.

Evolutionary psychology's metatheory is still evolving. Some evolutionary psychologists think, for instance, that leading researchers such as John Tooby and Leda Cosmides, Steven Pinker, and David Buss put too much stress on the universality of evolved human nature and too little on individual differences. Thus Patricia Gowaty (1992, 1997b, 2003) argues that the importance of within-species variation has been underestimated within the leading paradigm. According to Geoffrey Miller (2000a, 2000b) current metatheory is too restrictive, focusing too much on the survival value of the human mind and underestimating the role of sexual selection, which accounts much better for high individual variation. In the future the relative value of these different perspectives will have to be measured through multiple criteria, such as empirical testing, explanatory power, and predictive success.

Characteristics of Adaptations

The concept of adaptation has been carefully analyzed by evolutionary biologist George Williams in a now-classic work: *Adaptation and Natural Selection* (1966). An adaptation may be defined as an evolved solution to a specific problem that contributes either directly or indirectly to successful reproduction. Sweat glands, for instance, may be adaptations that help solve the survival problem of thermal regulation. Mate preferences may be adaptations that guide the successful selection of mates. Williams believed that the concept of adaptation should be invoked only when necessary and provided criteria for determining when we should do that (Buss 1999).

Indeed, in order to call something an adaptation a lot of criteria must be fulfilled. It must be a system of inherited and reliably developing properties that recurs among members of a species. It must be so well-organized and such a good engineering solution to an adaptive problem that a chance coordination between problem and solution is ruled out. Standards for recognizing adaptations thus include factors such as economy, efficiency, complexity, precision, specialization, and reliability, which render the design too good a solution to a defined adaptive problem to be a coincidence (Tooby and Cosmides 1992). This does not mean that it is a *perfect* solution. Selection favors a mechanism when the benefits outweigh the costs relative to other designs.

Looking for adaptations means distinguishing them from by-products of adaptations and from random effects or noise. The umbilical cord, for instance, is an adaptation. The belly-button that comes along with it is, however, a by-product. The particular shape of a person's belly-button may be considered as just a random effect (Buss 1999).

The standard example of an adaptation is the vertebrate eye. It consists of an extraordinarily complex, organized arrangement of specialized features, such as the pupil, the iris, the lens, the retina, the rods and cones in the retina, which transform light into electrochemical impulses, the optic nerve, and the visual cortex, where the incoming signals are further analyzed by a whole array of information-processing mechanisms that also constitute crucial parts of the visual system. Besides being highly coordinated and complex, the eye does something very useful for the organism: It provides information about the outside world. Through retaining those accidental modifications that improved performance, selection has gradually made the original light-sensitive patch of the original single-celled founder organism into the complex vertebrate eye (Humphrey 1992; Tooby and Cosmides 1992).

Ironically, this single most uncontroversial example of an adaptation is *as much* a physiological adaptation as a psychological one, as Tooby and Cosmides (1992) point out. It is designed to construct cognitive models of the world: of the objects that are present, of colors and shapes, of the faces that are present, and of the emotional expressions on those faces. The argument that perceptual mechanisms may well be adaptations but that the "higher" cognitive capabilities are not, since the first are evolutionarily older and hence have had more time to become highly elaborated, will not do. A lot of other adaptive problems have also been with us for tens of millions of years, such as the problems of reciprocation, maternal care, threat perception, mate selection, foraging, and emotional communication. Hence we have had the time to develop adaptive responses to them as well. Our language faculty, moreover, has the same overwhelming functional complex design as the visual system and yet it is a recent and human-specific adaptation.

Evolutionary psychologists do not automatically assume that, whenever a trait is found to be universal, it is an adaptation. In their classic study *Homicide*, for instance, Martin Daly and Margo Wilson point out that "[i]t is important to note that our evolutionary psychological approach in no way depends upon homicide *per se* being 'an adaptation'" (1988:12). As we saw, a number of criteria have to be fulfilled before a trait is thought to deserve that name. Furthermore this appellation has to be generally consistent with the findings from other disciplines.

With this consideration of adaptations we have touched upon a hotly debated issue: the degree of scientificity of evolutionary psychology.

Is Evolutionary Psychology Scientifically Defensible?

A critique frequently leveled at sociobiologists and evolutionary psychologists is that they are just engaging in after-the-fact-storytelling. They are producing unverifiable "just so" stories, because their claims are untestable and thus unfalsifiable. Evolutionary psychology, it is concluded, is not a science (e.g., Dupré 2001; Fausto-Sterling 1992; S. Rose 2000; Segal 1999; Tang-Martinez 1997). As Brian Haig and Russil Durrant (2000) note, the basis of this charge seems to be grounded in what critics see as the problematic nature of adaptationist explanations, because the concept of adaptation is necessarily a historical one. Since we can never directly observe natural selection in operation, we can never unequivocally demonstrate that any given characteristic really is an adaptation and hence amenable to an evolutionary explanation.

It must be acknowledged that just so stories *have* been told, such as Desmond Morris's unsubstantiated account of human female sexuality. Adaptive scenarios are, however, often mere heuristic models, ways to identify and study some of the many factors needed to fully understand the selective origin of traits suspected to be adaptations designed by selection. The use of these heuristic models is not the same as reaching conclusions (Waage and Gowaty 1997). Unfortunately some evolutionists have done exactly that, as many popularizers have. The reason that Morris's theory is no longer accepted today, however, is precisely because it has been empirically established as being untenable. This implies the point I want to make: Evolutionary explanations *are* falsifiable, and they are so at all levels of analysis, including the metatheoretical level.

A few examples. If mechanisms would be found in one species that are designed exclusively to promote the welfare of another species, the evolutionary metatheory would be falsified, since such mechanisms cannot arise through natural selection. They have not been found yet. If in any species the sex that invests more in offspring (usually the females) were in general less discriminating in choice of mates than the sex that invests less (usually the males), a key assumption within Trivers's parental investment theory would be falsified. All known species, however, fit the prediction (Ellis and Ketelaar 2000). The evolutionary hypothesis that men prefer attractive women because a woman's appearance provides a wealth of cues to her fertility can be falsified by testing whether the features that men consider attractive in women, such as youthfulness and a low waist-to-hip ratio, are linked to fertility. They are (Etcoff 1999).

Although sometimes it may take some creativity to design experiments, explanations in terms of adaptive function are generally testable, though not all to the same degree. One strategy for generating evolutionary hypotheses is called the observation-driven or bottom-up approach ('reverse engineering'): Researchers try to figure out the possible adaptive function of known phenomena, such as language, bipedal locomotion, pregnancy sickness, standards of physical attractiveness, or sexual jealousy. To do this, it must be shown that the phenomenon is well designed for solving a specific problem of survival or reproduction, and that it is not better explained as a by-product of another adaptation. This kind of analysis often leads to important new insights into the phenomenon and opens up new lines of investigation. A complementary strategy is the theory-driven or top-down approach: On the basis of evolutionary theory and knowing that our hominid ancestors lived as hunter-gatherers, we can make educated guesses about what design features our mind should have. These hypotheses can then be tested. If the hypothesized adaptations did evolve, they will manifest themselves as a reliable part of human nature across cultures, given the appropriate environmental input.

Evolutionary functional analysis thus provides a framework for the explanation of known facts and forms a powerful heuristic system for the generation of new knowledge. Although this may sound like a circular argument, it is not. Harvey asked why there were valves in the veins, and it led him to discover the circulation of the blood. Ever since, asking why a particular design exists has proven to be a very productive line of research in biology. It has led us to the prediction and subsequent finding of thousands of facts that were not known before (Tooby and Cosmides 1992). William Hamilton's formulation of inclusive fitness theory in the early 1960s, for instance, led to the discovery of psychological mechanisms producing behavior that conformed to this nonintuitive theory in many animal species. Knowing that our female forebears spent a lot of time gathering led Irwin Silverman and Marion Eals (1992) to predict that women should have evolved a superior object location memory, a prediction that was subsequently confirmed.

The "just so story"-charge can be countered in a second way: In modern philosophy of science, verification is generally considered too high a standard to be of practical use in evaluating scientific explanations.

It was philosopher of science Karl Popper who in his 1959 *The Logic of Scientific Discovery* explained that verification is untenable as a criterion for evaluating scientific statements. No amount of empirical evidence, he argued, actually *proves* that a particular explanation is true. Scientific statements can only be corroborated, that is, they can only be established as being consistent with the data. This is done by trying to falsify or refute them, on the basis of practical applications and experiments. From the set of nonfalsified (corroborated) alternative accounts, the best available explanation is then chosen (Ellis and Ketelaar 2000; Ketelaar and Ellis 2000).

Popper's concept of corroboration instead of verification has been retained in current philosophy of science. Indeed, strict adherence to verifiability as an evaluative criterion would eliminate, among others, archaeology, cosmology, evolutionary biology, forensic medicine, paleontology, tectonic geology, quantum physics, and all of the behavioral sciences as legitimate scientific enterprises (Ellis and Ketelaar 2000). This means that well-validated evolutionary explanations can count as acceptable knowledge claims.

The strict use of Popper's strategy of falsificationism, however, has been challenged by many philosophers of science as an inaccurate depiction of the way science actually works. Popper's contemporary Imre Lakatos, for instance, argued that in practice, scientists do not immediately reject a theory from the moment that a prediction derived from it has been falsified. This is because there are always different levels of explanation in science: There is a "hard core" of metatheoretical basic assumptions that are consensually held and not tested anymore, because they have been empirically established. These metatheoretical assumptions are the starting point for developing and testing more specific middle-level theoretical models. When finding anomalies here, scientists do not conclude that their metatheory has been falsified, but only that *this specific middle-level statement* has been falsified. Scientific theories are always evaluated by the cumulative weight of the evidence in relation to alternative theories. Scientists just try to create the best approximations of phenomena, based on evidential grounds and on certain agreed-on metatheoretical first principles, such as Newton's laws and Maxwell's laws in physics (Ketelaar and Ellis 2000).

Timothy Ketelaar and Bruce Ellis (2000) have analyzed evolutionary psychology from a Lakatosian perspective and have found its standards and procedures to be consistent with normal paradigm science. The three levels of analysis that Buss (1999) describes are consistent with the Lakatosian model of science, be it that instead of two there are three levels of explanation: the "hard core" of the general theory of evolution by natural selection, which includes inclusive fitness theory; middle-level evolutionary theories such as Trivers's theory of parental investment; and the level of specific evolutionary theories, such as the hypothesis that women have evolved specific preferences for men with a lot of resources. If predictions made at one level of the hierarchy are falsified, then the hypotheses on which they were based are called into question, but this does not in itself prove that the higher-level theory is wrong. The evaluation of evolutionary theories rests with the cumulative weight of the evidence, and not with any single prediction, which is just the way science works as described by Lakatos.

In evolutionary psychology, this evidence can come from a wealth of sources and can be obtained by a variety of methods. Potential sources of data for testing hypotheses are, for instance, archaeological records, data from hunter-gatherer societies, observations of people's behavior, self-reports, public records, and human products. Possible methods for testing hypotheses are to compare different species, to compare males and females, to compare individuals within a species, to compare the same individuals in different contexts, and the use of experimental methods. Again, it is the cumulative weight of the evidence that will be decisive in accepting a hypothesis (Buss 1999).

Another criterion of evaluation stressed by Lakatos and many other philosophers of science is a theory's ability to predict novel facts. Ellis and Ketelaar (2000) offer a list of thirty recent empirical discoveries about human psychology derived from the application of evolutionary principles to the human mind. They include sexually dimorphic mating strategies; waist-to-hip-ratio as a determinant of attractiveness judgments (Singh 1993, 2002); the use of cheater-detection procedures in social exchange; stepchild abuse at forty times the rate of nonstepchild abuse (Daly and Wilson 1985, 1988, 1995); sex-linked shifts in mate preferences across the lifespan; predictable patterns of spousal and same-sex homicide (Daly and Wilson 1988); maternal-fetal conflict during pregnancy; superior female spatial location memory (Silverman and Eals 1992); design of male sexual jealousy; profiles of sexual harassers and their victims; sex differences in the desire for sexual variety; socialization practices across cultures differing by sex and mating system (Low 1992); patterns of risk taking in intrasexual competition for mates; shifts in grandparental investment according to sex of grandparent and sex of parent; and mate guarding as a function of female reproductive value.

Besides predictive success and explanatory power, another potent criterion for evaluating scientific theories is their ability to mesh with other well-accepted theories and entrenched knowledge. Evolutionary psychology again meets this criterion effortlessly, as it does other criteria that scientists use as well to assess theories, such as their elegant simplicity, their internal coherence, and their ability to explain the underlying causal machinery at work (Fletcher 2000). We may thus conclude that the charge of unscientificity is not tenable.

Evolutionary psychology is frequently charged of being reductionist, whereby critics make reductionism sound like it is necessarily flawed (Benton 2000; Kimmel 2003; Rose and Rose 2000b; Segal 1999). This need not be the case, because actually there are different kinds of reductionism. In a positive sense, reductionism is just the way science works. It is now the virtually universally held position in science that (1) all things are physical things and nothing else (materialism), and that (2) the more complex and less fundamental can be explained in terms of the less complex and more fundamental. A human society consists of people who, being mammals, obey the rules of biology. Mammals consist of molecules obeying the laws of chemistry, which are, in turn, subject to the laws of physics (Dennett 1995; Plotkin 1997).

That does not mean that the more complex and less fundamental can be *reduced* to the less complex and more fundamental. This is reductionism in a negative sense, and it is this kind of reasoning that evolutionary psychology seems to be charged of—the critics seldom clarify what they actually mean by the term. Negative reductionism implies, for instance, that scientists would think that one does not need the levels of psychology and sociology in order to explain the rise of fascism or cubism: Molecules suffice (or, in this case, genes). No one, however, is a reductionist in this sense. Although we have seen that Wilson (1975) hinted that traditional sciences like psychology would eventually disappear, he has shifted his position markedly over a period of about a decade (Dennett 1995; Plotkin 1997).

An example of a confused interpretation of evolutionary accounts of the human mind is found in Segal (1999). "Accounts of 'loving' and 'working' direct us immediately toward quintessentially human narratives; 'surviving' and 'mating' strip away this human dimension," Segal complains (1999:80). She obviously conflates the causal forces that produced our psychology and that psychology itself, thinking that evolutionary scientists want to reduce our mental life to the forces that gave rise to it during our evolutionary past.

Explaining the ultimate origins of our subjective experiences, as evolutionary psychology does, obviously does not make these experiences any less real or any less valuable. Consistently thought through, Segal should level her critique at any research that regards humans as physical, biological beings (neuroscience, genetics, hormonal research, etc.), since these could also be said to "strip away the human dimension." Even the paradigms that she finds inspiring, Marxism and psychoanalysis, should be made her target, since humans are considered here as a product of certain historical or psychological laws. It should be clear, however, that doing so would be absurd. All these approaches just try to explain human functioning and behavior at yet another, complementary level of description.

Evolutionary psychologists are sometimes charged of being reductionists for another reason: because they fail to include the full range of variables that are needed to account for a given behavior. This is, however, the way science works: It tries to elucidate the nature of a phenomenon by simplifying its complexity and diversity into empirically manageable parts. If we would take on its full complexity as it appears in the real world, we could only offer a description and nothing more. We would not be able to generalize beyond the historical moment and the aspects involved. That this procedure of necessary reductionism works is demonstrated by the vast amount of knowledge produced by scientific disciplines (Campbell 2002; Thornhill and Palmer 2000).

Evolutionary psychology is producing a fast-extending body of knowledge on the human mind and human social behavior, such as on cognition, kinship relations, mating, cooperation, parenting, aggression and warfare, status striving and social hierarchies, and attachment. As my focus is on the evolved psychologies of women and men, I now turn to the ways in which evolutionary theorizing can be of use to feminism.

Notes

1. Robert Wright, for instance, writes that "Hrdy . . . may take a more than scientific interest in arguing that female primates tend to be 'highly competitive . . . sexually assertive individuals'" (1994b:69). It leads Hrdy to complain: "as if female primates really aren't, but owing to inner impulsions of my own I yearned them to be so" (1999a:xxviii). Still it must be added that Wright continues: "[t]hen again, male Darwinians may get a certain thrill from saying males are built for lifelong sex-a-thongs" (1994b:69), which shows him just being concerned for possible sources of bias in *both* sexes.
2. For example, Birkhead 2000; Ellis 1992; Gowaty 1992, 1997b, 1997c; Hrdy 1997, 1999a, 1999b; Low 2000; Malamuth 1996; Mealey 2000; Mesnick 1997; Miller 2000b; Shields and Shields 1983; Smuts 1995, 1996; Thornhill and Palmer 2000; Thornhill and Thornhill 1983; Waage 1997; Wilson and Mesnick 1997.
3. Oliver and Hyde 1993.
4. Bleier 1985; Blackman 1985; Fausto-Sterling 1992; French 1992; Haraway 1991; Harding 1985; Hubbard 1990; Lenington 1985; Noske 1989; Rose 1983; Rosser 1992; Sapiro 1985; Segal 1999; Tobach and Sunday 1985.

6

A Metatheory for Feminism

> [E]volutionary theory not only considers *how* men exercise power over women, as feminist theory does, but also investigates the deeper question of *why* males want power over females in the first place, which feminists tend to take as a given.
>
> Smuts 1995:2, emphasis in original

An environmentalist account of gender difference presents feminists with an insurmountable problem. As Josephine Donovan, reasoning within an environmentalist framework, recognizes: "[I]f gender identities and the cultures that go with them are historically a matter of social construction, then any political ideology based upon them . . . seems less secure than if based upon an immutable construct such as biology" (2000:76). Again it has to be added that Donovan is wrong in thinking of biology as "immutable." The false nature-nurture dichotomy that informs her thinking puts her in a difficult situation: *Either* it is a matter of "biology," which means nothing can be done about it, *or* it is a matter of cultural construction, which means feminism will have a hard time legitimizing itself. Indeed, cultural determinism easily slips into cultural relativism, with "culture" being used as a rationale for brutalizing women. What would, after all, be wrong with denying women education, sexual self-expression, or personal autonomy when their identity is just a cultural construct and they therefore have no inherent needs that can be ignored or violated? (See, e.g., issues raised by Susan Moller Okin [1999].)

Historically feminism has seen itself faced with similar perplexing questions. How to create a theory that allows for differences between women while at the same time justifying universal claims about them? Again the assumed polar opposition between 'biological determinism' and environmentalism blocks the possibility of finding any constructive way out. How to account for the fact that, despite many attempts to the opposite, girls and boys of each new generation start acting in sex-typical ways? How to explain the perhaps unsettling observation that "the available analyses suggest no consistent tendency for sex differences in social behavior and personality to have eroded or increased over time" (Eagly 1995:148)?

As we have seen, the nature-nurture dichotomy is thoroughly outdated. Any attempt to exhaustively explain human behavior in terms of nature only or nurture only is doomed to fail, as are attempts to account for gender differences in terms of cultural roles being "written upon" anatomical differences. Reality is much more complex and much more elegantly simple at the same time: Socialization is tremendously important, but it does not shape the development of behavior without guidance from innate dispositions. These dispositions, or potentials, have developed during the evolutionary history of our species. This means that males and females are not passive lumps of clay that culture can mold in any direction—an observation that can be empirically verified. They are active partitioners in their own development, acting in context-sensitive ways that were once adaptive to both sexes.

In this chapter I show that an evolutionary account of sex differences serves us with an answer to many questions that have divided feminism since its earliest days. I believe that evolutionary psychology can provide a unifying metatheory for feminism. That does not mean that the rationale of the existence of a multiplicity of feminisms is denied here. Evolutionary psychology, after all, only provides us with explanations; it does not prescribe how to go from here. It is up to us to choose to what extent we want to endorse our evolved predispositions and, as always in politics, points of view are expected to differ tremendously. The only way to make progress in feminism seems, however, to start from a scientifically defensible view of the nature of the sexes. Feminism is missing this—as are the social sciences in general. As Ellis and Ketelaar (2000) note, there are now about nineteen different grand theories of human nature in psychology, most of them making no precise or testable predictions, generating few laboratory studies, and frequently leading to disputes of opinion rather than fact. In this respect it is quite ironic that evolutionary psychology, which *does* meet these prerequisites, is considered by many as being unscientific.

The observations of male and female psychosexuality as made by feminists and evolutionary psychologists often converge, something that already indicates that the gap between both groups may be more imaginary than real. The difference is, however, that evolutionary psychologists tend to describe without indicting. As scientists, their job is to find out how things work. They try to provide the observed phenomena with an ultimate scientific explanation. Due to this attempt at neutrality, they are often considered sexist by feminists, but in fact this approach is their strength. As Simon Baron-Cohen says in an e-mail exchange with Lynne Segal, "[t]he training of a scientist is not to emphasize or package the results to make them politically correct, but to report on what is found. As scientists, we are judged by whether we are even-handed with the evidence" (Segal and Baron-Cohen 2003). Moreover, and not surprisingly, many evolutionists sympathize with feminist goals; knowledge of the evolved differences between the sexes promotes awareness of the costs of being a woman. Feminist socialization explanations can be valuable as accounts of the ways in which the innate behavioral predispositions of the sexes are culturally reinforced. They can help us to find ways of attenuating sex differences where we consider that a desirable goal. But feminists need a good underlying theory of human nature, and the modern theory of evolution by selection is exactly that. This is no intellectual imperialism. Evolution by selection is as sure a fact as is gravitation, and just like the laws of gravitation apply to all physical beings, so do the laws of evolution apply to all organic beings.

Evolutionary psychology does not want to do away with feminism. On the contrary, an evolutionary account of the sexes even attests to the importance for women to stand strong against men, because from an evolutionary point of view men are predicted to try to control female sexuality. Feminists need to get rid of their biophobia. As this chapter hopefully shows, biology can be our ally in the struggle for a more gender equal society.

Feminist Observations

Marilyn French (1992) notes that men are obsessed with female reproduction. "Because women bear children, men try to control or appropriate their bodies" (1992:18). The drive to control female reproduction, she says, is "a silent agenda in every level of male activity" (1992:19). She further presses charges against most states for trying to regulate sexuality by regulating *women*, for criminalizing prostitutes, although the consumers of prostitution are almost exclusively men, and for appropriating the right to determine whether women have access to contraception or abortion. Women, she points out, have been and still are being killed for losing virginity (even if they were raped) and for extramarital sex. They are put in purdah[1] or are infibulated. They are denied rights to divorce or to child custody. None of these constraints were ever imposed on men. Men have been killed for adultery in some societies, but only for having stolen another man's property—his wife—not for having sex with women other than their own wives. French interprets all these correct observations as signaling a conscious war against women, being waged because men do not want to take responsibility of babies.

Beauty is the main weapon in the competition between women, Susan Brownmiller (1984) observes. She further notes that, whereas being middle-aged does not detract from a man's perceived attractiveness, it does so for a woman. Behind this pattern she detects a male conspiracy to keep women weak and unsure by making them obsessed with their physical appearance. Joseph Lopreato and Timothy Crippen (1999) cite a study by Leslie Margolin and Lynn White (1987) on the differential role of physical attractiveness in marriage. Margolin and White note that men generally prefer mates who are younger and that this preferred age difference increases as men grow older. Their report provides considerable evidence from previous research that "beautiful women . . . trade their looks for economic status in cross-gender relationships" and that "as couples age, husbands are more likely than wives to lose sexual interest in their spouses." Margolin and White conclude that the "declining physical attractiveness of their partners affects marital sexuality more for men than women." Their explanation lies with "the cultural prescriptions and meanings associated with gender role behaviour." People's sexual preferences "result from an elaborate normative superstructure that informs perceptions and gives meaning to the things we do" (Margolin and White 1987, as cited in Lopreato and Crippen 1999:178–179).

Another correct observation is made by Valerie Bryson (1999), who points out that women are on average significantly less interested and active in formal politics than men are. Their involvement in less-institutionalized forms of collective activity, such as community groups and voluntary organizations may, however, be greater than men's. She further notes that feminists have found it extremely difficult to eliminate hierarchies, and that the role of leaders may actually be greater in nonhierarchical forms of organization than in more structured groups, because it cannot formally be checked.

Men have been found to be more violently aggressive than women (Hyde 1996). They are more likely than women to misperceive mere friendliness as a sign of sexual interest (Abbey et al. 1996). Almost half of all women scientists are concentrated in psychology or the life and social sciences. In contrast, a majority of men are in the field of engineering (Rosser 1992). All labor markets are highly segregated on gender lines, with women concentrated in caring, service, and clerical work and in the public sector, and men in manufacturing and industry and in the private sector (Bryson 1999).

The list of powerful feminist data continues. Brownmiller (1975) provides an overview of the prevalence of rape throughout history and indicts the unequal treatment of men and women before the law in rape cases. Correctly rejecting the then-predominating view of rape as a psychopathology, she substitutes it with the view of rape as an attempt to control women. In her notorious words, rape "is nothing more or less than a conscious process of intimidation by which *all* men keep *all* women in a state of fear" (1975:15, emphasis in original). Brownmiller's view of "rape as power, not sex" has since then been the reigning explanation within feminism (e.g., Abbey et al. 1996; Blackman 1985; Denmark and Friedman 1985; Dworkin 1997; French 1992; Hall 2000; Rich 2000[2]; Sanday 1986, 2003; Schwendinger and Schwendinger 1985; Travis 2003c; Watson-Brown 2000), although it is disputed by some other feminists (Hoff Sommers 1994; Roiphe 1993; Young 1999).

In an evaluation of some feminist claims regarding prostitution, philosopher Igor Primoratz (1999) acknowledges that a woman's history of sexual activity is not generally appreciated as an indicator of experience and expertise, analogously to other activities. He does not know the explanation for this, but "one is certainly needed" (1999:107). When analyzing the question whether the sexual sphere has a specific moral significance that sets it apart from other spheres of human behavior, Primoratz thinks the answer is negative. "What does apply to choices, acts, and practices in the field of sex are the same moral rules and principles that apply in non-sexual matters" (1999:173). In sex, just as in nonsexual matters, we can hurt, deceive, or exploit others, and we are morally required not to do so, he continues. This seems fair enough. His next statement, however, seems to be in need of qualification: "When we go against any of these requirements, the sexual nature of our conduct makes it neither more nor less wrong than it otherwise would be" (1999:173). Thus rape "is not wrong as *sexual* battery, but as sexual *battery*" (1999:174, emphasis in original). To sustain his argument, Primoratz refers to the feminist claim that rape is essentially a crime of violence, not of sex.

Intuitively there seems something wrong with Primoratz's evaluation of rape as being analogous to physical battery. We know that the psychological pain following rape is much greater than the psychological pain following physical battery (Thornhill and Thornhill 1990). Why should that be, if the sexual nature of the act does not make it any different? Primoratz recognizes that something does not fit here, and suggests that it might be the intimate nature of the sexual act that makes it distinctive in some weaker sense. But, as he argues, that cannot be it, because sexual relationships are not the only ones that are experienced as intimate, nor are they always experienced as such. Thus his conclusion remains the same, and the reader is left wondering what it is, indeed, that makes rape such an undeniably horrible crime to women.

Another frequent complaint is that, although there is some increase in men's efforts in the household and in childcare, "studies in every country consistently find that 'new man' is more of a myth than a reality, and that the traditional domestic division of labor remains remarkably unchanged, even when women are in paid employment" (Bryson 1999:124). Many women like staying at home and taking care of the children. Even in the Scandinavian countries, where both sexes can enjoy family friendly working arrangements, men are much slower to make use of them than women (Bryson 1999).

Men are also much more interested in visual sexual stimuli than women and are the main consumers of pornography and prostitution (Angier 1999). According to radical feminists, pornography has this appeal to men because it mediates and helps maintain the sexual subordination of women, through its eroticization of male power and female submission. By presenting women as objects to be bought and used, pornography shapes people's worldviews and cultivates a view about sexual relationships as power relationships (Dines, Jensen, and Russo 1998; Dworkin 1997; French 1992).

There is a long list of systematic, well-established feminist and sociological observations about male-female relations, as well as an accompanying list of less well-established and often contradictory explanations of these observations. Other examples include typical courtship rituals, age difference at first marriage, differential rates of marriage and remarriage, differential maternal and paternal relations to children, and differential rates of infidelity (Lopreato and Crippen 1999). There is not, however, within feminism nor within the social sciences, a powerful, coherent, and well-founded theory which can explain these patterns. Referring to cultural prescriptions is no exhaustive explanation, since that leaves the existence of those prescriptions unexplained. Moreover, as we have seen, social learning explanations fail to account for these observations even if one just takes the existence of cultural rules for granted.

The Seeds of Adulthood

As discussed in the preceding chapter, the evidence indicates that human infants are born prewired and genetically disposed to attend to specific components of the environment, to act in certain ways, and to have some specific preferences. The further examples I offer in this section are particularly interesting in that some of them show unexpected sex differences. That is, unexpected—and unexplainable—from an environmentalist framework, not from an evolutionary one.

Infants stare at human faces intently. Less than ten minutes after birth their eyes will follow facial patterns. On the second day they can distinguish their mother's face from a face they have never seen before. Soon they start mimicking facial expressions. They gaze almost as long at a person's eyes as at the face as a whole. When they see somebody looking at them, babies typically look back and smile (Etcoff 1999). Newborn infants automatically focus themselves on the things they need to survive. They are born to attach themselves passionately to their primary caregiver—typically the mother—and to manipulate her into taking care of them by displaying evolved characteristics such as crying and being cute. They are no blank slates, but possess a range of dispositions that are structured to be activated by certain environmental stimuli. This goes from simple actions such as gripping and sucking, over more complex behaviors such as looking for facelike patterns, preferring melodical, high-pitched voices, and, around six months of age worldwide, acquiring fear of strangers, to very complex cognitive capabilities such as discerning the content of speech sounds (Hrdy 1999b).

Another partially prewired design appears to be our sense of physical beauty. Contrary to traditional psychological theories of attraction, which assume that standards of attractiveness are learned gradually through cultural transmission and hence will not emerge before three years of age, standards of beauty have been proven to emerge very early in life. When three-month-old infants were shown slides of white and colored faces that had been evaluated for their attractiveness by adults, they gazed significantly longer at the more attractive faces. Universally infants have other sensual preferences as well. They prefer symmetric patterns to asymmetric ones and soft surfaces to hard ones. At the age of four months they prefer consonant melodies to dissonant music (Etcoff 1999).

Simon Baron-Cohen (2003) describes a series of intriguing experiments that reveal that from birth, girls are more people centered than boys, whereas boys are more object centered. One-year-old infants were filmed while they were playing on the floor and their mothers sat nearby in a chair. The girls looked up significantly more at their mother's face than the boys did. When given two films to watch, one of a face and one of cars, the girls looked longer at the face, whereas the boys looked longer at the cars. These differential preferences were also found in babies who were just one day old. They were filmed while a girl came smiling over their crib, and while a mobile came hovering over it. The mobile was made from a ball the size of a head, with the face of the girl projected onto it; only, its features were rearranged so that it no longer looked like a face. Some material was attached to it that moved every time the mobile moved. The researchers, who did not know the babies' sex, found that girls looked longer at the face, whereas boys looked longer at the mechanical object. Also, at the age of twelve months, girls respond more empathically to other people's distress, through sad looks, sympathetic sounds, and comforting behavior. By the age of three, girls are ahead of boys in judging what other people might be feeling or intending, that is, in using a theory of mind (Baron-Cohen 2003).

This sex difference in social interest, with girls and women on average being more empathic and communal than boys and men, is well-known in the psychological literature (Eagly 1995). Socialization cannot account for the finding that this difference already shows on the first day of life.

Moreover, in contemporary Western nations child-rearing practices have been found to be very similar for both sexes. The few exceptions are that parents tend to give their children sex-appropriate toys, with boys being discouraged from playing with dolls, especially by their fathers, that boys receive more punishment and physical discipline, that they are more allowed to horse around, and that mothers talk more and provide more emotional speech to their daughters than to their sons (Baron-Cohen 2003; Campbell 2002; Geary 1998). Rather than parents' sex-typed reactions creating differences in the child's preferences and inclinations, however, they might actually *stem* from differences that are already present. Remember from chapter 5 that sex-typical toy preferences emerge very early in life, even without specific encouragement. Boys are generally far more attracted to playing with mechanical objects than to playing with dolls. When they do prefer dolls and other typical girls' toys, chances are high that as adolescents they will develop a homosexual orientation (Bailey 2003). This correlation appears to be at the basis of fathers' negative response to feminine playing preferences in a son. From as early as two years old, boys are more dominant, more aggressive, and more physically active than girls, which might account for their greater freedom of movement as well as for their more frequent disciplining. Infant girls' better empathizing, in turn, might cause mothers to attune their speech to this emotional capability (Baron-Cohen 2003; Campbell 2002; Geary 1998).

Again, this is not to say that parenting and socialization are not important. They are crucially important, but not in the way that many people think they are. Psychologist Judith Rich Harris demonstrated in her notorious book *The Nurture Assumption* (1998) that culture and parenting mainly seem to provide the conditions that allow for normal development and happiness. They do not determine personality, nor do they create the differences between girls and boys. They merely mediate the expression of these features. They provide a newborn baby with the necessary stimuli and opportunities for the skeletal competencies of its evolved modules to develop themselves in ways that make the child fine tuned to the local ecology. The often-found correlations between the personality of parents and children do not attest to the molding influence of child-rearing, as has often been uncritically assumed, but appear to be due mainly to genetic influences (Ridley 2003).

Conclusive evidence that biology plays a part in girls' and women's relatively greater focus on people and in boys' and men's relatively greater focus on objects came when Baron-Cohen (2003) and his students found a relationship between prenatal testosterone levels and the development of these characteristics. During the prenatal period, from approximately eight to twenty-four weeks, the levels of testosterone are much higher in males than in females.[3] Differential amounts of exposure to sex hormones have been shown to have different organizational effects on the brain, with typical cognitive sex differences as a result (Kimura 1999, 2002). For the first time now, prenatal hormonal exposure has also been linked to sex differences of a deeply *social* nature. After correcting for all social variables, Baron-Cohen (2003) found that the lower a baby's prenatal testosterone, the more eye contact it will make as a toddler and the larger its vocabulary will be. At four years of age, lower levels of prenatal testosterone will have led to better levels of language, better social skills, more eye contact, and better empathizing skills. The higher its prenatal testosterone, the more restricted its interests will be and the more focused and better it will be at systemizing, such as playing with technical objects, building Lego castles, throwing and intercepting balls, and collecting and classifying items.

The typical and cross-cultural patterns of children's play, with boys engaging more in rough-and-tumble play and group-level competition and girls engaging more in play parenting, have also been found to be influenced by prenatal hormone levels (Geary 1998). Again this implies that socialization explanations will not suffice to explain behavioral sex differences. Rather, socialization and parenting practices seem to amplify or attenuate the innate dispositions of girls and boys. In an analysis of ninety-three cultures, Low (1989) has shown that cross-cultural differences in the way both sexes are reared depend in predictable ways on group size, marriage system, and degree of stratification. Predictable, that is, again, from an evolutionary perspective: Boys and girls are reared in ways that enhance their social, economic, and reproductive success in that specific social system. Thus, the more polygynous the society, the more sons are taught to display aggression and industriousness, because of the big variance in reproductive success: A very few men will be able to acquire many wives, whereas many men will find no wife at all. In stratified polygynous societies where women can "marry upward," daughters will be trained more strongly to show sexual restraint and obedience, because this increases a girl's mate value to prospective high-status husbands. The more control women have over resources, on the other hand, such as in matrilineal societies, the less daughters will be taught to be submissive and obedient.

Childhood differences between the sexes thus appear to result from a complex interplay between genetic, hormonal, developmental, and social factors. The long time that human children take to mature corresponds to the high level of social and cognitive sophistication of our species. This extended period of immaturity, made possible by prolonged parental care, allows them to develop their evolved competencies in accordance with the local environment. In this way they can acquire the knowledge and experiences necessary for successful survival and reproduction. Hence we can expect girls and boys to be differentially oriented to developing the competencies that are associated with intrasexual competition in adulthood, such as risky behavior and dominance-striving in boys and physical enhancement and the verbal derogation of competitors in girls. We can also expect girls to be much more oriented toward parenting behavior than boys (Geary 1998).

Sexual Selection as an Origin Theory

Consistent with a sexual selection framework, the sex differences found in childhood become more pronounced at puberty, when a new surge of sex hormones modifies the physiology, cognition, and behavior of both sexes. Secondary sexual characteristics emerge. Males develop larger hearts and lungs, greater upper-body strength, longer legs, higher running speeds, and better throwing skills—all features that would have been important for the tasks of hunting and fighting. The male advantage in throwing skills is already found at two years of age and is related to a difference in the skeletal structure of boys that emerges in the womb (Geary 1998). Hence it cannot result merely from boys' greater practice at throwing, as Fausto-Sterling (1992) has suggested.

Girls enter puberty two years earlier than boys. Their hips widen and their body fat increases relatively more, with a typically feminine distribution pattern. They show greater physical flexibility and develop a finer motor dexterity. The physical size difference between the sexes and the faster sexual maturation of females are typical of polygynous primates. Comparative studies have revealed a dependable relation between the degree of physical dimorphism and that of male-male competition in these primate species. The harder males have to compete for sexual access, the higher the selection pressures for size and strength, because it will often be the larger and stronger males who are preferred by females and are able to spread their genes and, hence, their traits (Geary 1998).

Men's and women's minds differ, too. This difference is one of degree, not of kind. As Alice Eagly describes, feminists have long tried to deny or minimize these differences, but they can no longer look past "the onslaught of contemporary research documenting sex-differentiated behavior" (1995:150), although this continues to be done in most contemporary psychology textbooks. The fairly intuitive methods that were typically used for synthesizing psychological studies on gender a few decades ago have now been replaced by statistically justified procedures. The results they yield are solid and unanimous: Instead of being negligible, there are moderate-to-large sex differences in some aspects of cognitive abilities, personality and social behavior, sexual behavior, and physical abilities (Eagly 1995). Females typically score higher on verbal fluency, spatial location memory (remembering the location of objects), and mathematical calculation, for instance. Males typically score higher on mental rotation tests (trying to imagine what a given figure will look like in a new orientation) and mathematical reasoning (Eagly 1995; Kimura 1999, 2002). The fact that girls and boys outstrip each other in different aspects of mathematical and spatial ability suggests that teachers' expectations cannot account for these findings, since most teachers are not even aware of these differences. Moreover, these differential abilities are found cross-culturally. The difference is not necessarily greater in more traditional societies, such as Japan, again suggesting that socialization fails as an exhaustive explanation (Kimura 1999; see her book for further discussion of why socialization explanations are inadequate).

When it comes to personality and social behavior, women are typically more tenderminded, nurturing, and socially sensitive, whereas men are more dominant, controlling, and independent. These descriptions fit the stereotypes, but then again, gender stereotypes have been found to be an *underestimation* of real gender differences, rather than an exaggeration (Eagly 1995). A large body of scientific research further shows that men's and women's sexuality differs in four important respects. First, men have more interest in sex than women, as exemplified by such facts as that they think about it more often, that they report more frequent sexual fantasies and more frequent feelings of sexual desire, that they rate the strength of their own sex drive higher, masturbate more often, and are more interested in visual sexual stimuli. Second, aggression is more strongly linked to sexuality for men than for women, with rape as an extreme manifestation of this difference. Third, women tend to place greater emphasis on committed relationships as a context for sexuality. Fourth, women's sexuality is more plastic than men's: It is more easily influenced by environmental factors and more likely to change over time (see, e.g., Peplau 2003; Peplau and Garnets 2000; Schmitt et al. 2003a, 2003b).

Many feminist researchers "have worked energetically to preserve the 1970s scientific consensus that sex-related differences are null or small," Eagly (1995:150) writes. Their research "was (and is) intended to shatter stereotypes about women's characteristics and change people's attitudes by proving that women and men are essentially equivalent in their personalities, behavioral tendencies, and intellectual abilities" (Eagly 1995:149). It is this attitude that characterizes the work of critics like Anne Fausto-Sterling and Ruth Hubbard.

Their attempts were in vain, however. Instead of being false constructions, gender stereotypes have turned out to be fairly accurate representations of typically male and female traits—something that should not surprise us, as Eagly (1995) notes, since we all interact with members of both sexes on a daily basis. So the question now no longer is *whether* women and men differ, but where these differences come from.

As elaborated in chapter 4, no single theory of socialization succeeds in providing a basic or ultimate explanation of gender difference. Some prior knowledge about the existing social structure is always needed in order to be able to make predictions about male and female psychology and behavior.

At first sight, one seeming exception might be social role theory, as proposed by Alice Eagly and Wendy Wood (Eagly and Wood 1999; Wood and Eagly 2000). Social role theory is different from other socialization theories in that it is, like evolutionary theory, a 'theory of origins' (Archer 1996): It searches for a basic cause of sex-differentiated behavior. Social role theorists trace gender difference back to biology, but not in the sense of heritable, evolved psychological dispositions, as evolutionary psychologists do. Rather, they suggest that physical sex differences, in particular men's greater size and strength and women's childbearing and lactation, lead to a predictable pattern of sexual division of labor, because certain activities are more efficiently accomplished by one sex. This in turn creates a sex-typical psychology and sex-typical competencies, as men and women try to accommodate to the gender roles that society has created for them. "Sex differences in behavior thus reflect contemporaneous social conditions" (Eagly and Wood 1999:414).

How probable is social role theory *as an origin theory*? Extremely improbable, as should be clear by now. The finding that prenatal testosterone levels are a crucial determinant of social behavior (Baron-Cohen 2003) suffices to refute the theory. Formulated more generally, social roles cannot account for the finding that variations in the sex-differentiated organizing influence of prenatal androgen levels on the brain results in corresponding variations in sex-typical inclinations and competencies. They cannot account for the finding that in men and women, natural changes in hormonal levels (across the menstrual cycle for women; across the seasons and within the course of the day for men) are associated with predictable changes in cognitive and physical strengths, such as verbal fluency, spatial abilities, and motor dexterity either (Kimura 1999). Social role theory cannot explain that the pattern of sex differences that characterizes humans is typically found in other mammals as well. Its presupposition that human behavior has not been (partially) sculpted by sexual selection is contradicted by the earlier-mentioned relation between a species' degree of physical dimorphism and its degree of intrasexual competition between males. As a matter of fact, social role theory does not at all fit in with the broader framework of all that is known in the biological and life sciences.

Apart from this devastating finding, the theory can be tested by comparing the predictions that it generates with those generated by evolutionary psychology. As Eagly herself says, "[a]ll theories of sex differences have the task of explaining not just overall differences in various behaviors but also the patterning of these differences across studies and therefore across social settings" (1995:149). In the next sections on mate preferences we see that an evolutionary perspective does a far better job at explaining the findings than social role theory does.

The only theory of origins that does not run into trouble when confronted with the scientific evidence about gender difference is sexual selection theory. From a natural selection perspective, the sexes are expected to be more or less alike, since they were confronted with the same problems of survival, such as finding food, keeping warm, and avoiding predators.[4] Their problems of reproduction, however, were disparate. Because of their unequal extent of parental investment, women and men could enhance their reproductive success in different ways. As a result they are expected to show predictable psychosexual differences. This does not imply that they, consciously or unconsciously, *wanted* to reproduce as much as possible. As Tooby and Cosmides say, "[i]ndividual organisms are best thought of as adaptation-executers rather than as fitness-maximizers" (1992:54). It just means that individuals with characteristics and mate preferences that made them outreproduce others, would have spread those traits within the next generations. As the offspring of ancestors who managed to reproduce successfully, men and women today are expected to have inherited their traits and preferences.

From a reproductive perspective, not everybody is equally valuable as a sexual partner. In evolutionary jargon individuals whose characteristics make them potentially have a large and healthy progeny are said to be high in mate value. From this perspective, young women have a higher mate value than older women, since older women's fertility and reproductive potential are much lower and after a certain age even decline to zero. Men's mate value, on the other hand, will not be as much affected by age as women's, since men's fertility declines much less steeply with age. Ancestral men who had psychological mechanisms that made them feel more sexually attracted to older than to younger women would not have left as much progeny as individuals with opposite preferences. The latter would have outreproduced them, with the male preference for youthful mates ultimately becoming species typical. This evolutionary logic leads us to expect that feelings of sexual and romantic attraction will be typically associated with traits in the other sex that entailed high reproductive success in the world in which our ancestors evolved (Ellis 1992).

The sexual choices exerted by members of the opposite sex are called sexual selection pressures: The partner preferences of men help direct the course of the evolution of female physiology and psychology, since they influence which female characteristics will be able to spread over the next generations. The same holds for women's partner choices. Apart from this *intersexual* selection, there is also *intrasexual* selection, but both are linked. Intrasexual selection, or intrasexual competition, means that males should evolve over time to compete with each other to obtain those resources and display those qualities that females prefer. The more they do this, the more they will be able to satisfy their own evolved mate preferences—sometimes to the detriment of women, as we will see. Females as well should evolve to compete with each other to display those cues that males find attractive, since for them, too, the degree to which they can get what they want depends on their own mate value (Buss 1992).

The following sections briefly sketch how the qualities that women and men typically desire in a mate are differentiated by sex, and how the patterns of difference that are found between male and female psychology fit a Darwinian framework.[5] A critical distinction that will be made in analyzing women's and men's mate preferences is whether they are seeking brief affairs or long-term mating partners, because typically both sexes are expected to adjust their standards accordingly (Buss and Schmitt 1993). When reading this account of male and female psychology, it is important to keep in mind that it concerns a *continuum* of difference, not a dichotomy, and that cultures can and do mediate the expression of these evolved inclinations to varying degrees. Gender ideology obviously influences individual behavior; it is only the initial direction of these traits which is given. As Smuts (1996) says, gender ideology itself can be viewed as a reflection of the reproductive interests of the most powerful individuals in a given society. When women resist patriarchy and get more power, they will bend gender ideology toward ways that empowers the female half of the population, as we have seen happening in the West. Hence Hilary Rose's (1997) conclusion that if evolutionary psychology were true then there is no point in feminists working for a society that accepts diversity but resists domination is clearly false.

An Evolved Male Psychology

For men more than women, reproductive success is constrained by sexual access to fertile mates. Women are, as Trivers (1972) called it, a 'limiting resource' for men: As a result of their greater parental investment there will always be a smaller number of reproductively available women around than there are reproductively available men, and as a result of women's smaller reproductive potential they will be more picky than men about whom to mate with.

There is thus an initial asymmetry between women and men on the sexual playing field. Given this, plus the observation that human males provide substantial parental care, plus the observation that concealed ovulation and internal fertilization make paternity uncertain for men but not for women, a wealth of predictions about male (and female) mating psychology can be generated.

Men should, for instance, have evolved psychological mechanisms that make them feel sexually attracted to features that signal youth and health and hence fertility or future reproductive capability in the other sex.[6] These features are hypothesized to constitute what we call physical beauty. Men should also value virginity and sexual fidelity in a potential long-term mate more than women do, because this increases the probability that her offspring are also his. They should have psychological adaptations that help guarantee their paternity (Buss 1992). They can furthermore be expected to compete vigorously among themselves for sexual access to the more investing sex. They should be more risk taking, more violently aggressive, and less empathic than women, since more is at stake for them in the intrasexual struggle. Whereas every woman, being a member of the high-investing sex, can likely find a man willing to mate with her, not all men will be in that luxurious position. They will have to compete for female attention, particularly in those domains that women find important, otherwise they might not find a mate at all (Trivers 1972). They should also typically pursue a mixed sexual strategy, longing for a long-term relationship as well as having a relatively greater desire for casual sex and brief affairs than women, since both strategies are reliable ways of enhancing male reproductive success (Buss and Schmitt 1993; Trivers 1972).

Major studies for assessing these predictions are Buss's (1989) cross-cultural study on mate preferences, and the International Sexuality Description Project (Schmitt et al. 2003a, 2003b, 2004). Furthermore there are many psychological, sociological, and evolutionarily inspired studies on these subjects. In general they provide support for the evolutionary approach. We zoom in on a number of findings.

The male preference for beautiful and young women is so blatant that it is well-known to everybody. It is a cross-cultural phenomenon, too: Everywhere men more than women rate beauty as important in a potential mate, and they typically desire someone younger than themselves (Buss 1989). The structural constituents of what we experience as female beauty are consistent across cultures as well and have been established as a reliable cue to fertility. Female physical attractiveness amounts to signs of youth and health. Smooth and clear skin, full lips, lustrous hair, white teeth, a graceful gait, firm breasts: All are cues to high fecundity. Beauty is also constituted by symmetry and by having average features relative to the rest of the population. Both symmetry and being average are positively linked with health and a good genetical constitution in the whole of nature. Furthermore, attractive female faces have proportions that signal a combination of sexual maturity and relative youth. As a result of sexual selection, we also experience as beautiful those characteristics that magnify indications of femininity: large eyes, high cheekbones, full lips, and a small chin. Women worldwide use makeup to accentuate these feminine features and to make themselves look younger (Buss 1992; Cunningham et al. 1995; Etcoff 1999; Symons 1995). Another (near-)universal standard of female beauty appears to be a 0.7 waist-to-hip-ratio (WHR), which is positively linked with health and fertility.[7] The findings are complicated, however, because WHR interacts with Body Mass Index: Whenever body weight becomes very high or very low, it seems to affect men's ratings of female attractiveness more than does WHR (Singh 1993, 2002). More research on the precise interaction between the many variables affecting perceptions of female beauty is needed.

The cultural expression of standards of beauty may vary. In cultures where food is scarce, plumpness signals wealth and health, and thus indicates high social status. In cultures where food is abundant, the wealthy distinguish themselves by being thin, and as a result thinness becomes the cultural ideal. The cultural variation in ideals of female body weight is thus not arbitrary and is not likely to be greatly affected by exposure to Western media. African American men, for instance, continue to express greater attraction to a heavier build in women than white American men, a preference that, according to social psychologist Michael Cunningham and his colleagues, likely reflects enduring uncertainty concerning resource availability (Cunningham et al. 1995). What remains stable across time and across cultures, however, is the male preference for a low WHR. From the upper Paleolithic over ancient Greek, Indian, and Egyptian cultures until today, female WHR is depicted most frequently in the range of 0.6 to 0.7 (Singh 2002). The fact that most Upper Paleolithic Venus figurines have a 0.7 WHR but are very fat suggests that early humans recognized the link between a critical amount of body fat and fertility (Hrdy 1999b).

Of course men need not be, and indeed are not, aware that their experience of sexual attraction has an evolutionary rationale. They do not have to know why smooth skin and a narrow waist appeal more to them than wrinkles and other signs of old age in order to typically find the first traits more attractive than the latter. Underpinning this taste, however, are psychological adaptations.

Buss (1989) found that cross-culturally, the characteristics most desired in a long-term mate are similar for both sexes: kindness and intelligence. Men as well as women prefer as a long-term mate an intelligent and understanding individual who is compatible with them. As expected, however, youth and beauty occupy a much more central place in male than in female mate preferences.[8] The results of this survey in thirty-seven cultures differing widely in ecology, location, ethnic composition, religious and political system, and mating system show other persistent and predictable sex differences as well. Men indeed value virginity in women more than women do in men, although cultures vary tremendously in the value placed on it. Factors explaining this variability are, among others, the economic independence of women, the prevailing incidence of premarital sex, and the reliability with which chastity can be evaluated. The first two factors interact: When women control their economic fate and hence are less dependent on men, they are freer to disregard men's preferences. This causes the average amount of premarital sex to go up (Buss 1989).

The valuation of female chastity fits in the worldwide pattern of men's typical obsession with female sexual fidelity. The universality of male sexual proprietariness, which has engaged feminists for long, is not surprising from an evolutionary point of view: It is likely to evolve in any species with internal fertilization and paternal care, and the parental investment made by human males is quite significant. By monopolizing a woman they minimize her opportunity to cuckold them. They thus reduce the risk of investing time and energy into children not their own while at the same time they could have been procreating elsewhere. Men do not consciously think that way, of course, nor do all men exhibit this tendency. The point is that ancestral males who did not care about a mate's promiscuity or about seizing other sexual opportunities themselves, only had a small chance of transmitting their genes and, thus, of actually becoming our ancestors. They were outreproduced by males who regarded their sexual partners as a kind of property and who were able to make a trade-off between time invested in mate guarding, in parenting offspring, and in trying to sire more offspring elsewhere. As a result a sexually proprietary psychology became typical of many human males (Wilson and Daly 1992).

Psychological adaptations to reduce the risk of cuckoldry have been found in other paternally investing male animals as well. In male birds, for instance, copulation frequency, intensity of mate guarding, and provisioning of young have been shown to vary adaptively in relation to cues to possible extrapair copulations by the female, such as her period of fertility, the mating strategies of rivals, the degree of coloniality, his attractiveness relative to rivals, and lapses of his surveillance of the female (Wilson and Daly 1992). Human males' tactics of mate guarding and their degree of sexual jealousy, too, have been shown to vary according to cues to their partner's mate value, their own perceived mate value, and possible signs of the partner's unfaithfulness—since, as we see in the next section, females have evolved to look for short-term mates as well. As expected, sexual jealousy in women has been found to be qualitatively different, triggered more by cues to a mate's emotional infidelity than by his sheer sexual infidelity. Maternity is certain for her, but when her mate becomes emotionally involved with another woman, he may channel part of his resources and attentions to that other person instead of to her and her children (Buss 2000). As said before, this does not imply that all men experience sexual jealousy, only that many men will, and that on average they will experience it more intensely than women do. It is just as with body size: The fact that some women are larger than some men does not detract from the observation that men are, on average, the larger sex.

The combination of male proprietariness, a female tendency for promiscuity (see next section), and greater male aggressiveness gives rise to a dangerous cocktail. Indeed, worldwide male acts of violence against women have been shown to be mainly inspired by sexual jealousy. Men sometimes assault their partners in response to actual or suspected sexual infidelity or as a way of deterring them from desertion. The risk of physical abuse and even lethal violence is proportionate to a woman's mate value: Young and attractive women run a much higher risk of being assaulted by their mate than elderly or unattractive women. This is particularly the case when the woman is much younger than her partner, since, being young, she is in a good position of finding a more desirable mate. Stalking as well is mainly done by men, and most stalkers have had a relationship with their victim, a relationship that she ended (Buss 2000; Wilson, Daly, and Scheib 1997). The patterning of these acts of violence, as well as the general fact that worldwide it is *young* men who are most aggressive and commit most of the violent crimes, is hard to explain in terms of social role theory, but it fits sexual selection theory like a glove: In ancestral environments young males wanting to find a mate had to prove their mate value by displaying their physical prowess and the ability to defend their interests. In this way they could impress women as well as rival males, cultivating a reputation of bravery (Archer 1996; Buss 1999; Daly and Wilson 1988).

Male sexual proprietariness has more unsavory manifestations, such as the development of a gender ideology and social customs and structures that are meant to control female sexuality. The degree to which men succeed in doing this will depend on many factors, such as their control of resources (see section on the origins of patriarchy). This core male mind-set results in well-known phenomena that are, to use the words of Margo Wilson and Martin Daly, "culturally diverse in their details but monotonously alike in the abstract" (1992:291): phenomena like veiling, purdah, claustration, infibulation, clitoridectomy, chaperoning, the valuation of female chastity, the concept of women as property and of adultery and rape as a property violation, the denigration of prostitutes, the equation of the protection of women with protection from sexual contact, and laws that excuse male violence that was provoked by "loose" behavior of a wife, daughter, or sister. It is women of reproductive age who are most restricted in their freedom, not children or postmenopausal women. The rationale for this pattern is, again, hard to identify from a social role perspective, but obvious from a Darwinian point of view. This does not mean that women are doomed to having their sexualities suppressed by men, since male coercive control is conditional. If female resistance becomes too organized and hence too costly to fight, men will be forced to tune down (Smuts 1996).

Because they are the less-investing sex, men are predicted to have evolved a greater desire for sexual variety than women, to direct their short-time mating effort most intensely toward women who are sexually accessible, to let less time elapse before seeking sexual intercourse, to relax their standards in short-term mating, and to avoid commitment when pursuing brief sexual encounters (Buss and Schmitt 1993). Indeed, men have been found to be less exacting than women when it comes to casual sex. This seems to be a psychological adaptation to securing a large number of casual sex partners. Age, intelligence, kindness, personality, and marital status become much less important. Beauty, however, becomes even more desirable in a short-term context. In contrast to their long-term preferences, men like cues to promiscuity and sexual experience in a potential temporary mate, probably because these characteristics increase the likelihood that they can gain sexual access. In contrast to the tremendous positive value that men place on commitment when looking for a marriage partner, for casual sex they strongly dislike women seeking a commitment. They rate only four negative characteristics as significantly more undesirable than women do: a low sex drive, physical unattractiveness, a need for commitment, and hairiness (Buss 1994; Buss and Schmitt 1993).

Numerous other studies have confirmed these predictions about male desires and male behavior. In the International Sexuality Description Project, a recent extensive survey across fifty-two nations and ten major world regions by psychologist David Schmitt and colleagues, for instance, sex differences in the desire for casual sex turned out to be culturally universal (Schmitt et al. 2003a). The survey provided support for the theory that both sexes have evolved to pursue long-term relationships as well as brief affairs, depending on the circumstances, but that men will look for casual sex more often and more intensively than women. This theory is known as sexual strategies theory (Buss and Schmitt 1993). Across every time interval, ranging from "in the next month" to "in the next thirty years," men were found to desire significantly more sex partners than women within every world region. Moreover, even when women reported to be strongly seeking a short-term mate, they typically did not want more than one sex partner in the next month. Men who were strongly seeking a short-term relationship, on the other hand, tended to desire more than one sex partner in the next month, with the number of desired partners strongly increasing over time. For women the increase was only marginal over time. Women were found to be slower in consenting to sex than men, and to be less actively seeking for casual sex (Schmitt et al. 2003a).

Of course, self-reports have their limits. Women and men may bias their answers in directions that they experience as socially desirable. Thus, men may exaggerate the number of sex partners they have had, whereas women may tend to present themselves as more chaste than they really are. Indeed, even after controlling for visits to prostitutes and for homosexual orientation,[9] men turn out to on average report larger numbers of past sexual partners than women do (Volscho and Pietrzak 2002). This is statistically improbable; the total number of sex partners should be equivalent for both sexes (but not necessarily for each individual man and woman; many men may have had more sex partners than many women if a number of women are highly promiscuous). This finding suggests that some of the sex differences in self-reports of sexuality result from differential reporting rather than actual differences in behavior. In a recent experiment, psychologists Michele Alexander and Terri Fisher garnered evidence that in sexual self-reports women feel compelled by social mores to distort the true level of their sexual experience, a pressure not so much felt by men (Alexander and Fisher 2003). Participants were asked to complete a questionnaire under three different conditions: a bogus pipeline condition in which they thought lying could be detected, an anonymous condition, and a nonanonymous condition. When anonymity was not guaranteed, women reported a smaller number of past sexual partners. It was this part of the survey that received a lot of media attention ("women lie about sex"), but the essential point of the study was more than this, as Terri Fisher herself noted (personal communication). Women's self-reports about *some* aspects of their sexual *behavior* indeed turned out to be false accommodation to gender norms. Their sexual *attitudes*, however, did not seem to be really influenced by social expectations. The finding about biased self-reports of behavior finally redresses the statistical anomality of a differential number of sex partners for both sexes, as well as the previous finding of an earlier age of first intercourse for boys than for girls—a finding that, if it were true, would mean that adolescent girls have first intercourse with a boy younger than themselves, whereas they typically prefer older boys. No significant sex differences were found in sexual permissivity across all three testing conditions. Other sex differences remained significant across the testing conditions: differences in erotophilia (positive emotions toward sex), in frequency of masturbation, and in exposure to pornography. Moreover, the size of the sex difference across the anonymous and bogus pipeline conditions was very small, suggesting that anonymous testing conditions may be as valid as bogus pipeline conditions. As Alexander and Fisher note, however, most sex surveys are conducted in settings where participants' identities risk to be revealed.

The results of the experiment lead them to suggest that

> sexual behaviors may be more susceptible to social desirability responding and self-presentation strategies than are sexual attitudes. If this is the case, findings on sex differences in self-reported sexual attitudes may indicate real differences between the sexes whereas the typical patterns found in self-reported sexual behavior may not accurately reflect true sex differences. (Alexander and Fisher 2003:33)

The complexity of Alexander and Fisher's findings was not only overlooked in the press. An example of selective reading can be found in a reaction from liberal feminist Cathy Young, who picks the finding about the lying women and contrasts it with Schmitt and colleagues' (2003a) findings about the greater male desire for sexual variety worldwide (Young 2003). Caricaturing evolutionary psychology as "the notion that men are 'hard-wired' by evolution to spread their seed while women are predisposed to seek monogamous relationships," Young paints a misleading picture of both studies. Her selective way of presenting them makes their respective results look as if they were incompatible, with Schmitt and colleagues being exposed as fabricating "sex stereotypes that are the stuff of late-night comedy routines." People who read both studies carefully will have to acknowledge that Young renders their results in a highly biased and selective way.

Back to male psychology now. Social role theory, as presented by Eagly and Wood (1999), predicts that when women gain socioeconomic equality with men, their sexuality will become similar to men's. This prediction is refuted by Schmitt and colleagues' findings. Since this survey tried to maximize anonymity as best as possible, and since Alexander and Fisher's (2003) bogus pipeline research showed that answers given in anonymous conditions differ only slightly than under bogus pipeline conditions, we may consider its results reasonably trustworthy.

Moreover, these results converge with many other findings about a distinct male and female sexual psychology, such as the finding on sex differences in quantity and quality of sexual fantasies. Men's sexual fantasies are twice as frequent as women's, and they involve more anonymous and multiple partners. They specify many more sexual acts and a greater variety of visual content, and are more sexually explicit than women's (Ellis and Symons 1990). The study by Ellis and Symons reports no sex difference in the (high) degree of enjoyment of sexual fantasies, nor in the experiencing of sexual desire for and the fantasizing about tabooed partners (such as one's best friend's lover or one's in-law). This suggests that sex-typical fantasy patterns are not a mere product of gender ideology. If they were, women would experience more guilt about fantasizing than men, since girls' sexual activities are typically circumscribed more than boys'. Moreover, as Ellis and Symons write, if they were, women would not be as likely as men to violate social conventions by admitting experiencing and enjoying forbidden desires and fantasies.

Pornography provides another illustration to the case. Male-oriented pornography is a combination of all that titillates male sexual psychology in the context of short-term relationships. The story is straightforward, directed at sex as sheer physical gratification, and women are willing sex objects, making no emotional demands (Ellis and Symons 1990; Symons, Salmon, and Ellis 1997).

An Evolved Female Psychology

Women are the more-investing sex and men provide a relatively high level of paternal investment. As a result, we can expect humans to be characterized by female-female competition and male choice, in addition to male-male competition and female choice. A Darwinian point of view generally predicts women to have psychological adaptations that give rise to sexual and romantic attraction to those traits in men that would have helped their children to survive in the evolutionary past: traits that provide cues to health, resources, protection, and his willingness to engage in parenting. Because paternal investment is reproductively beneficial for women, they can be expected to compete among themselves for men exhibiting these traits (Ellis 1992; Trivers 1972). Having a father around generally means more food and protection, reduced mortality risk, higher social competence and competitiveness, and higher future social and cultural success for a child. Male investment is thus important, but not absolutely necessary for child survival. This observation, together with the fact that men can never be sure of paternity, implies that evolutionarily it would have benefited men more than women to shift between parenting and mating effort. Men and women today can still be expected to differ in this respect, with women more than men investing time and energy in parenting relative to looking for short-term affairs. This relative difference in preferences is expected to be a source of conflict between the sexes, since women will typically try to obtain more paternal investment than men prefer to give. Variability between men will be high, however, with some men investing most of their time in parenting effort and others in mating effort, since both strategies are reproductively viable. Variability between women will be lower, since because of their greater initial parental investment they have less to gain in pursuing additional mates (Buss and Schmitt 1993; Geary 2000).

People are, of course, not consciously calculating the possible reproductive benefits of the choices they make in life. They just act on their preferences. It is instructive to remember the example of sugar: We do not have to know that foods rich in sugar helped our ancestors to survive and reproduce in order to like its sweet taste.

The finding that worldwide, across subsistence activities and social ideologies, mothers do the bulk of childcare and are far more frequently involved in spontaneous interaction with their children than fathers attests to their differential priorities. The same pattern is even present in the Israeli kibbutzim, in liberal Sweden, and in nontraditional Western families, suggesting that it may be something hard to change (Campbell 2002).

Why girls and women should extend their caring attitude to their social relationships in general, stressing intimacy, sensitivity, and interdependency more than boys and men do, is not completely clear. It may follow indirectly from their maternal investment, in the sense that it is important for child survival to create a stable and nonhostile social atmosphere, as Geary (1998) suggests. Certainly in the face of patrilocality, when women had to immigrate into a community of strangers at adolescence, it was crucial for them to establish friendly relationships. Campbell (2002) thinks it more likely that creating bonds of trust might have helped to protect women from male abuse. Both hypotheses are compatible, and so is the proposition that male-male competition and the importance of social dominance for male reproductive success would have prevented the evolution of high empathizing and of the need for emotional self-disclosure in men (Baron-Cohen 2003).

Female mothering has long been taken for granted by male evolutionary biologists, Hrdy (1999b) writes. They viewed motherhood as an uncomplicated task and consequently supposed that our prehistoric female ancestors produced as many children as their reproductive span would allow. Since a hunter-gatherer way of life allows for approximately one child every four years, each woman would have sired and raised her maximum of five or six children. Due to this presumed scarcity of selection pressures on women, they were not thought to benefit from competition, hierarchical striving, or promiscuity. What would they compete for, if they reached their maximum of reproductive success anyhow? We now know, however, that among hunter-gatherers women show considerable differences in their ability to produce children *and* to keep them alive. Many women die without leaving even one surviving child behind.

This variability in the reproductive success of female animals in general, as well as the impressive variation in mothering strategies that they display, was only discovered in the last decades of the twentieth century. It is now acknowledged that keeping her offspring alive requires intense planning and strategizing on the part of the female. Depending on the species, she will have to make a lot of choices: whom to select as a mate, how to disguise promiscuity, whether or not to abort, how to space births, when to stop weaning, how to prevent infanticide by a hostile male, how best to divide her time and energy between her offspring, how to balance between protecting them and stimulating their independency, how to combine taking care of her own needs and of theirs, or how to find alloparents—that is, other individuals who will help her look after her offspring. Many species of bird, fish, reptile, and mammal even reveal the baffling sophistication of facultatively investing differentially in sons and daughters, favoring the sex that will probably have most reproductive success in that particular context. They can do this by influencing the sex ratio before fertilization, by selective abortion, or by treating male and female offspring differently. When the circumstances change, their pattern of sex-biased investment shifts accordingly (Hrdy 1999b).

Some of these strategies will sound familiar. Humans are, of course, not just like any other animal species, but, as Hrdy (1999b) says, the fact that sex-biased parental investment pervades the animal world suggests that in the human case we cannot just attribute this phenomenon to gender ideology without simultaneously looking for the roots of this ideology.

From a Darwinian point of view women, and female primates in general, are not expected to be as violently aggressive and risk taking as males (except when defending their offspring), since they do not have to compete for copulations per se, and since their death considerably raises the hazards of their offspring dying. This does not mean, however, that they should have evolved to be passive and noncompetitive. Dominance pays for females, too, albeit not to the same degree as for males. Not in the way of sex, but in the way of access to food. In most social mammals, female reproductive success increases with female social status. High-ranking mothers introduce their daughters in their social networks, with their daughters inheriting the benefits of their mothers' rank: access to the best food, earlier sexual maturity, higher chances of offspring survival, and less harassment of herself and her offspring by other females. In primates the intensity of female-female competition depends on the abundance and spacing of the food. This appears to be the reason why female bonobos can establish close bonds: They do not have to compete, because food is so plentiful (Campbell 2002).

Nonhuman female primates compete directly for the resources they need to survive and reproduce. In humans these resources have often been monopolized by males, and consequently women will have to compete for them indirectly: by competing for men possessing resources. Moreover, since human males provide a high degree of parental investment, women will also compete for men exhibiting traits that promise paternal involvement. We see that among evolutionists there is disagreement on the origins of the female preference for wealthy men, as well as on the nature of women's sexuality.

When considering women's mate choices, I again distinguish between long-term and short-term mating contexts. Although, as elaborated in chapter 4, females have long been considered the monogamous sex, we now know that in many species, including humans, this is not the case. Women as well as men have mixed sexual strategies within their repertoire, be it that women are hypothesized to have evolved a lesser desire for sexual variety than men and to impose higher standards for short-term mates than their male counterparts (Buss and Schmitt 1993).

When it comes to long-term relationships, women as well as men regard love as an indispensable ingredient, although women value this cue to commitment still a bit higher than men do. With love thus being a prerequisite for a long-term relationship for both sexes, and kindness and intelligence as the most valued traits in a long-term mate for both sexes, women and men are very similar with regard to their major priorities in this context. The order of the items next on their list of qualities desired in a prospective long-term partner, however, differs significantly.

Dozens of studies document that American women value good financial prospects in a mate roughly twice as highly as men do, a difference that has remained constant over the last seventy years. In this preference for resources they follow the females of many other species. As David Buss writes, "[t]he evolution of the female preference for males offering resources may be the most ancient and pervasive basis for female choice in the animal kingdom" (1999:104). Hundreds of studies worldwide, ranging from cross-cultural surveys to analyses of personal ads in newspapers and magazines, have shown that women care a lot about a man's provisioning ability. Universally women also prefer high social status and ambition in a mate, probably because these are cues to his present or future control of resources (Buss 1989, 1994; Ellis 1992; Low 2000; Mealey 2000).

An obvious objection is that, since women lack socioeconomic power, they are forced to seek men with money (Angier 1999). This explanation implies, however, that whenever women get access to power and wealth, they should lay less stress on a man's earning capacity, across and within cultures, and that men who have less socioeconomic power should value financial resources more in a prospective mate. The hypothesis is thus testable and it is contradicted by nearly all available data. Across and within cultures, women's mate preferences become *more*, rather than less, discriminatory as their wealth, power, and social status increases, a finding that fits evolutionary theory, since the higher a woman's mate value, the more she can demand in a mate. Powerful women want superpowerful men. Women want partners whose socioeconomic level at least equals theirs, regardless of how high that level is. The preferences of men with fewer resources are, moreover, indistinguishable from those of wealthy men (Buss 1989, 1994; Buss and Schmitt 1993; Ellis 1992).

However, when Eagly and Wood reanalyzed Buss's data on mate preferences in thirty-seven cultures by combining them with measures of gender equality in these cultures, they found a weak trend in the opposite direction. Cross-culturally women indeed emphasize a mate's earning potential more than men do, but they tend to decrease this emphasis as gender equality increases. Nonetheless they keep ranking it as highly important (Eagly and Wood 1999). Holding feminist values has also been found to attenuate this preference to some extent, although women with feminist values keep adhering to the pattern of wanting wealth and status in a mate (Koyama, McGain, and Hill 2004).

It cannot just be that women may feel they have less job security than men, as Angier (1999) suggests. Among the Bakweri in Cameroon, the women are wealthier and more powerful than the men, because women are in much scarcer supply—an imbalance that results from the continual influx of men from other areas to work on the plantations. Still they insist on men with money or resources (Buss 1994). In hunter-gatherer societies, where resources are not easily accumulated and where women's gathering is as vital as men's hunting, the social status of men is an important consideration in women's mate choices. In societies like these, a man's status is defined by his sociopolitical activities, his hunting skills, his athletic prowess, or his knowledge. A high social status substantially contributes to his reproductive success: High-status men have more wives, more extramarital affairs, and more children, with their children having a higher chance of survival (Campbell 2002; Geary 1998). This female predilection for male status even in societies where women are not disadvantaged socioeconomically suggests that it might have very deep roots, but it might be too early to know that for sure (see also the section on the evolutionary origins of patriarchy).

The finding that cross-culturally women prefer as long-term partners men who are older than they are (three and a half years on average) is probably related to this female preference for status and wealth in a male. Typically a man's age provides a cue to his access to resources, since worldwide social status and wealth tend to accumulate with age (Buss 1989, 1994). Other important determinants of women's mate choice are a man's height, strength, and athletic ability. Some have suggested that these traits are of importance to women not only because they contribute to a man's status—they would have made him a proficient hunter, for instance—but also because they imply that he will be able to protect them against physical and sexual domination by other men (Smuts 1995, 1996; Mesnick 1997; Wilson and Mesnick 1997). Tall and athletic men indeed have more dates than short, average, or less athletic men (Etcoff 1999), which in times predating birth control would have implied more offspring.

For women physical attractiveness in a long-term mate should be important as well, but not to the same degree as for men, because of the significance of cues to paternal investment. Indeed, as in many species of the animal world, people of *both* sexes place high value on indicators of a mate's good health, such as bodily and facial symmetry (Etcoff 1999; Symons 1995), but women do so to a lesser extent than men (Buss 1989). Moreover, the components of male physical attractiveness are more complex, since women typically perceive cues to a man's social dominance as contributing to his attractiveness. A somewhat taller than average body size, personality characteristics such as self-confidence, independency, and assertiveness, and physiognomic traits stressing masculinity, such as prominent eyebrows, deep-set eyes, and a broad jaw, are all linked to social dominance and social status in men and, consequently, perceived as sexually attractive by women (Ellis 1992; Etcoff 1999).

Women are, however, not only looking for socially dominant men, but also for men willing to invest in children. As documented by Buss (1989), they most of all want a mate who is kind and understanding. Since personality traits enhancing social dominance do not always contribute to a man's kindness, we may expect this female preference for cues to dominance to be moderated by signs of a nice character. Indeed, when traits signaling dominance behavior and masculinity become too outspoken, such as having a very square jaw, their perceived sexual attractiveness diminishes. Women's sexual preferences, however, shift across the cycle. At midcycle women rate the sexual attractiveness of masculine facial features at its highest, and during that phase their preference for the scent of symmetry is at its strongest. This finding implies that women are looking for genetic quality when they are most likely to conceive. Moreover, their libido increases, as well as the probability of their having extra-pair copulations—the best father might not always be the one having the best genetic constitution (Thornhill and Palmer 2000).

Men do not show a specific preference for female willingness to invest in children: Experiments have revealed that, whereas women's ratings of the attractiveness of men as long-term mates are affected by cues to their affection or indifference toward a child, men's judgments are not. This is apparently because other than men, women do not vary widely in their affection toward children, so men did not "have to" evolve a specific predilection for these qualities in a long-term mate (Buss 1999).

A person's mate choice will thus depend on an intricate mixture of variables. Its precise manifestation will moreover be influenced by many other factors, including her or his personality, age, and life history, and, in many cultures, pressures from kin (who are, in turn, pursuing their own reproductive benefits). This interaction of variables explains why Darwinians do not generally expect young women to be drawn to substantially older men. A man's status may typically accumulate with age, but many women are also looking for strength, athletic skills, health, physical attractiveness, and compatibility. Nevertheless, the combination of male and female preferences sometimes leads to the familiar phenomenon of a fifty-five-year-old male rock star or captain of industry having a relationship with a twenty-two-year-old fashion model. Both have attributes that are highly valuable to the opposite sex; hence, they are in a position to actualize their mate preferences. She has traded her youth and beauty—typical weapons of intrasexual female competition—for a high-status male, whereas he has used his social status and resources—typical weapons of intrasexual male competition—for getting a highly desirable young woman. Both have employed their high mate value for acquiring a trophy mate.

In evolutionary biology men have long been considered the promiscuous sex, whereas women were assumed to be tending toward monogamy. This is rather unlogical; men could hardly have been selected for promiscuity without polyandrous women around, except in the context of rape. But because of societal prejudice, and because the possible adaptive benefits of female promiscuity were not obvious, this female tendency has long been overlooked. Meanwhile several possible benefits have been advanced, but they have not yet been subjected to many empirical tests. An ancestral woman might engage in casual sex in exchange for meat, goods, or services. In addition, by obscuring paternity, she might elicit resources from several men (Hrdy 1999a, 1999b). Another possible resource is protection (Smuts 1995, 1996). She might also engage in a short-term affair because of its genetic benefits: if her regular mate is infertile or impotent, in order to get different genes, or in order to get better genes. Among other possible benefits we find manipulating her regular mate into increasing his commitment and the possibility of finding a better mate (Buss 1999).

The costs of promiscuity are, however, predicted to be higher for a woman than for a man. She risks physical and sexual abuse, as well as reputational damage. This damage will likely be more severe than for men, ultimately because of paternity uncertainty: Men will not favor cues to promiscuity in a prospective long-term mate, since chances are that she will cuckold them. Another reason is that people may interpret a woman's promiscuity as a sign of her low mate value, since women of high mate value are generally discriminating: They know that they can afford to be choosy, because they are desired by many men. This trend occurs even in relatively promiscuous cultures such as the Ache, a foraging people in Paraguay (Buss and Schmitt 1993).

Sexual strategies theory proposes that women will not lower their standards as much as men do when seeking short-term mates. Whereas men can be expected to sometimes seek sexual variety per se, since this would have been of reproductive benefit to our male ancestors, women are not predicted to gain from an indiscriminate strategy. Just mating around with any willing man would likely have entailed too many costs for our female ancestors. They should have focused on those males who were likely to provide them with some benefits, be it resources, protection, good genes, or the prospect of a (better) long-term relationship. Indeed, Buss and Schmitt (1993) have tested these predictions and found them confirmed by their own research as well as by many other studies. Women frequently expect and obtain resources in exchange for short-term sex, with voluntary prostitution as an obvious example. The authors cite from an extensive cross-cultural analysis of prostitution, conducted by Nancy Burley and Richard Symanski (1981), who concluded that "the motives expressed by many sorts of women in numerous societies are, at their core, clearly economic: men create a demand and women find an economic advantage meeting it" (cited in Buss and Schmitt 1993:220). Buss and Schmitt also found that women place greater value on cues to immediate resource extraction in short-term than in long-term mates, despite the fact that they are generally less demanding in a short-term than in a long-term mating context. These findings support the resource hypothesis about women's short-term mating. Next, women more than men dislike characteristics in a short-term mate that indicate bad long-term prospects: when that person already has a relationship or is promiscuous. Moreover, they keep looking for love and emotional intimacy in short-term relationships—both findings that support the mate switching or long-term mate finding hypothesis. Women also place greater value on a man's strength and physical attractiveness in a short-term than in a long-term context, which backs up the good genes hypothesis, since attractive features generally connote a history of healthy development (Buss 1999; Buss and Schmitt 1993).

Sexual strategies theory further predicts that these differential strategies and preferences of both sexes will often be a cause of conflict. Men will be more likely to complain about women's sexual withholding, whereas women will be more likely to feel disturbed by men's sexual aggressiveness. Many studies have documented this effect (Buss 1999; Buss and Schmitt 1993).

That women keep up high standards in many contexts is reflected in their sexual fantasies, as well as in female-oriented erotic literature. Women's sexual fantasies are built up slowly and are more elaborate than men's. Women, more than men, emphasize touching, feeling, and the responses of their imagined partner. They are more likely to fantasize about someone they are involved with, and to focus on personal characteristics of that person. They are much less likely than men to switch partners during the course of a fantasy, and more often imagine themselves as recipients of sexual activity (Ellis and Symons 1990).

Erotic romance novels, which are almost exclusively written by and for women, provide another window to female psychosexuality. This immensely popular product differs profoundly from male-oriented pornography. Here the love story is the central action. Sex is important, but it does not dominate the plot. Romance novels are about a woman's desire to find the one right, passionate man who will remain hers for the rest of her life, not about her pursuit of sexual variety. Sexual activity is described primarily through her emotions rather than through descriptions of her physical responses or through visual imagery (Ellis and Symons 1990).

Hoff Sommers (1994) presents a typical example of this genre:

> Townsfolk called him devil. For dark and enigmatic Juan, Earl of Ravenwood, was a man with a legendary temper and a first wife whose mysterious death was not forgotten. . . . Now countrybred Sophy Dorring is about to become Ravenwood's new bride. Drawn to his masculine strength and the glitter of desire that burned in his emerald eyes, the tawny-haired lass had her own reasons for agreeing to a marriage of convenience. . . . Sophy Dorring intended to teach the devil to love.[10]

Are romance novels an attempt to indoctrinate women? It seems not. The free market is mainly driven by consumer choice, and sales are a measure of public preference. Romance novels, in which almost every hero is handsome, strong, and socially dominant, amount to almost 40 percent of all mass-market paperback sales. Attempts by feminists to have this literature replaced by a new one have failed. The readers' resistance to politically correct romances featuring sensitive, unaggressive heroes and sexually experienced, right-thinking heroines was too strong (Hoff Sommers 1994).

We might conclude that, whereas in men's sexual fantasies the goal is sexual gratification, women's fantasies tend to be related to mate choice, something that again supports the mate-switching or mate-finding hypothesis.

Thus three of the four earlier-mentioned gender differences in sexuality turn out to be predictable from an evolutionary perspective: men's greater interest in sex, the stronger link between aggression and sexuality for men than for women, and women's greater emphasis on committed relationships as a context for sexuality. The fourth difference, women's greater sexual plasticity (Peplau 2003; Peplau and Garnets 2000), is harder to explain. Women's sexuality is more malleable and capable of change over time. It is more responsive to social contexts, and women are more likely than men to have bisexual feelings or to change their sexual orientation across the lifespan.

Men are more likely to be either heterosexual or homosexual and to stay that way (Bailey 2003). Anthropologist Helen Fisher (1999) thinks that women's sexual flexibility results from an evolutionary history where it was crucial for a woman to secure the help of whoever was present and willing to be an alloparent to her children. Their greater sensitivity to environmental circumstances might also have served their reproductive interests—until patriarchal coercive techniques put a spoke in their wheels. It seems that currently we know too little about this aspect of female sexuality to say much about it. As psychologists Linda Garnets and Anne Letitia Peplau (2000) contend, women's sexuality and orientation has too long been conceptualized in terms of male experiences, resulting in ignorance about the complexity of this aspect of women's lives.

An indication that women's sexuality is more associated with environmental circumstances than men's was also documented by Schmitt and colleagues in their International Sexuality Description Project (2003b). They found that worldwide both sexes shift their styles of romantic attachment depending on several sociocultural characteristics, but that women do so more than men. Particularly, women tend to adopt the more typically male style of avoiding close personal relationships in environments with higher mortality rates, fewer resources, and higher fertility rates. This makes evolutionary sense. When resources are scarce and mortality is high, a woman might do better by shifting from a primary long-term strategy to a more short-term mating strategy with lower levels of emotional commitment. Early reproduction allows family members to help raise offspring, she may get resources and protection from more than one putative father, and she may get sexual access to men with good genes that can make her children withstand the harsh environment.

The perplexing result, however, is that gender differences in degree of emotional attachment are *most* marked in Europe and America, cultures with the most progressive gender ideologies and the highest degree of socioeconomic equality between the sexes. In Africa and Asia, cultures with the most traditional gender roles, women and men are very alike in their avoidance of emotionally close romantic relationships. The degree of this gender difference was found to be unrelated to cultural masculinity levels. These findings run directly counter to Eagly and Wood's (1999) social role hypothesis, which predicts that gender differences should decrease as women gain socioeconomic equality. With respect to attachment styles they *increase* instead: Women become more nurturing and more in need of emotional closeness (Schmitt et al. 2003b).

What should we infer from this finding about women's sexuality? It gives more weight to Barbara Smuts's and Sarah Hrdy's suggestion that patriarchal arrangements might have mutilated an originally assertive female sexuality, whether in the course of evolution (Hrdy 1997) or in the course of individual women's lives (Smuts 1996). As Smuts points out, women still grow up in a sociocultural context that represses the development and expression of their sexuality. This context may foster a female sexuality that responds to male needs and desires rather than having desires of its own. She thinks it is premature to attribute the relative lack of female interest in sexual variety to women's inherent sexual nature in the face of this repression and calls for suspension of judgment until we have more evidence.

Both are right in calling for caution, but we must not forget that there are a number of factors that unambiguously point to evolved differences in psychosexuality either. The finding that the magnitude of sex differences in attachment style is more strongly linked to evolutionary pressures such as a harsh ecological environment, mortality rates, and local sex ratio,[11] than it is to measures of gender equality, for instance. Or the finding that there are no sex differences in feelings of guilt over sexual fantasies, although the content of these fantasies differs considerably. Or the finding that testosterone administration to postmenopausal women increases their sexual fantasy and sexual activity, and that estrogen and antiandrogen administration to male-to-female transsexuals decreases their sex drive (Bailey 2003). The link between sex differences in sexual behavior, attitudes, and fantasies, and the sex differences in exposure to sex hormones, does not just result from social roles and expectations.

Benefits of an Evolutionary Framework

Evolutionary psychology does not want to usurp other disciplines. It only wants to *and can* provide an overarching theoretical framework that can organize and explain the observations done by feminists and social scientists. It is no competing theory to psychological and sociological theories; it is a metatheory. It provides us with a much-needed underlying scientific view of human nature. Women researchers are involved in it to nearly the same extent as are men, so any inkling of male bias will be severely scrutinized. Evolutionists do not at all deny that socialization affects human behavior and the expression of sex differences, but they argue that it informs only part of the story. They offer a broader, and complementary, explanation.

Evolutionary psychology does not imply genetic determinism. It is not reductionist in the negative sense, nor is it unscientific, inherently sexist, or any of the other things it has been charged of being. It fits in with what we know about the world and how it works. The preponderance of evidence speaks for it. It accounts for as many phenomena as those described by social scientists, and for many more. It is true that some of its hypotheses are at the moment not much more than this: just plausible hypotheses (e.g., Geoffrey Miller's 2000b account of the role of sexual selection in shaping the human mind). But they just await further research.

So what are feminists waiting for? Why not seize this powerful theory to sustain feminist claims? Feminist sociologist Amanda Rees acknowledges that "a self-aware, or reflexive, evolutionary psychology could have an interesting role to play in feminist theory" (2000:368). She summons her fellow feminists to start paying more attention to what evolutionary psychology has to say. Indeed, many of the phenomena mentioned in the section on feminist observations are not only explained by evolutionary theory, but actually *predicted* by it. The existence of conflict between the sexes, as well as the specific factors that provoke it, are to be expected given the differential sexual psychology of women and men. Some might object that this reveals evolutionary psychology to be determinist after all, but that is a misconception. As said before, the theory holds that the particulars of the manifestation of psychological adaptations will depend on the context. Our evolved dispositions are triggered by environmental cues, and changes in the environment will lead to changes in the way they express themselves.

If you are never thwarted in your objectives, you may never get angry. If your mate never flirts with others, you may never get jealous. If you grow up in complete isolation, you will never learn how to talk and even your sexual appetite may not quite develop. When confronted with the evolutionarily expected stimuli (stimuli that reliably manifested themselves in the course of evolution), however, such as a linguistic and socially complex environment, or, later in life, a mate who may be contemplating to leave you for someone else, we will respond in a species-typical way that is at the same time tailored to the specific context. If you are born in France, your language module will get you to spontaneously and effortlessly acquire French; in Germany it will be German. If you are having a romantic relationship, the probability that jealousy will manifest itself will depend on many factors, including your gender, your partner's behavior, your partner's mate value, your own mate value, the quality of your relationship, and your experiences earlier in life. Because of our evolved modules we can react in ways that come naturally and spontaneously to us, but that actually involve very complex computations. The more we get to know about these underlying mechanisms of behavior, the more we can learn about how to control it.

People who would never think about questioning the legitimacy of psychological or sociological research sometimes ask me what evolutionary insights are good for. I continue to be amazed by this. Is it not tremendously important to try to understand human nature? I suspect that the main reason is the false dichotomy between nature and nurture that is still preeminent in people's minds: It is good to study "nurture," since that implies we can find ways of remedying human behavior, but why should we study "nature," that fixed entity we cannot change? And aren't we cultural beings who have left the constraints of nature behind us? As I hope to have shown, things are more complex than that.

There are more possible answers to give. Many people do not seem to realize that the recognition of a human nature is a prerequisite for the foundation of firm ethical standards. As Steven Pinker (1997:48) asserts, "[i]f people's stated desires were just some kind of erasable inscription or reprogrammable brainwashing, any atrocity could be justified." A denial of human nature, no less than an emphasis on it, can be used for harmful ends.

Pinker describes a 1974 documentary about the war in Vietnam, where an American officer explains that we cannot apply our moral standards to the Vietnamese. Their culture, he says, does not place a value on individual lives, so that they do not suffer as we do when family members are killed. The director plays the quote over pictures of wailing mourners at the funeral of a Vietnamese casualty, thus refuting the officer's rationalization. Individual rights, Pinker points out, can only be founded by assuming that people have intrinsic wants and needs and are authorities on what those wants and needs are. I would add that the same holds for women's rights. Evolutionary insights can thus serve us with some guidelines in, for instance, the debate on multiculturalism and gender equality.

Evolutionary psychology is also an explicit refutation of the suggestion that one sex might be superior to the other. As David Buss (1996) explains, any notion of superiority is logically incoherent from an evolutionary point of view. A bird's wings cannot be considered inferior or superior to a fish's fins, and in the same way neither sex can be considered inferior or superior. Each sex possesses mechanisms designed to deal with its own adaptive problems. Most of these mechanisms are similar for women and men; some, specifically in those domains having to do directly or indirectly with reproduction, are different. Evolutionary theory carries no normative values; it is descriptive.

Moreover, evolutionary psychology does not consider women any less capable than men of being rational, of reasoning in an abstract way, or of using general moral principles, as some difference feminists and ecofeminists do—a feminist claim that is, in fact, rather insulting to women.

Indeed, it is actually environmentalist accounts that might be considered quite sexist, since women are regarded as weak and willing, as helpless victims in the hands of men. An evolutionary perspective focuses on female choice as one of the primary mechanisms driving gender difference. It highlights a female mind that has an identity of its own, instead of merely being the product of patriarchal oppression. Consequently it offers a more affirmative view of women. As Douglas Kenrick, Melanie Trost, and Virgil Sheets (1996) point out, for instance, the phenomenon of trophy wives, in which younger, attractive females have a relationship with older, powerful males, has often been explained by social scientists as a product of cultural pressures. In this view women are passive victims manipulated and traded by all-powerful men. As the authors argue, however, the very name "trophy wives" illustrates an unwitting sexist bias common in environmental accounts of gender difference: the tendency to ignore the role of female choice in mating relationships. Similar patterns of age difference and of women trading their youth and attractiveness for power in a mate are found throughout the world and throughout history and are very likely due to sexual selection. As explained before, it makes evolutionary sense that older males should seek relatively younger females, whereas the trophies that females seek consist of social status and resources in a mate, both factors that typically accumulate with a man's age. Following evolutionary insights a much more complex picture emerges. It is one where males and females are active players in the game and where their different mate preferences and reproductive strategies have ultimately led to the rise of patriarchy (see section on the origins of patriarchy).

As I argued in chapter 2, I do not think that scientific theories should be accepted or rejected on ideological grounds. We should not defend an evolutionary approach to the sexes simply because it puts women in a better light than men (the latter emerging as the more violently aggressive, more sexually proprietary, more status-striving, and less nurturing sex). The elegant truth is, however, that evolutionary psychology both accounts better for the available data and gets much more empirical support than does environmentalism. Feminists may have trouble interpreting this view of female psychology as positive, since women's lesser extent of status striving and dominance behavior does not promise for their rapid advancing to the higher regions of power. I would retort that we finally have a theoretical framework showing us that typically female traits and abilities are as important as typically male ones. I would add that, as feminists, we should not get trapped by what so many male theorists have done before us: to take the male mind and male behavior as our standard. Moreover, if we can formulate good reasons for why we think there should be an equal number of women and men at the top, such as the proposition that a government ought to represent the whole of the population, we might choose to introduce policies to achieve that goal. Evolution does not dictate to us how to live our lives.

Evolutionary psychology empowers women on a theoretical level, but on a more personal level as well: It shows them that they can be as feminine or as nonfeminine as they feel like. They do not have to fear that their "wanting to feel like a woman" betrays their weakness of character. They can be just whoever they are: feminine or not, heterosexual or not, nurturing or not. While evolutionary psychology is about the universal design of the human mind (and of that of the sexes), it is also about variation. Without variation, after all, there can be no evolution.

Taking an evolutionary approach can further help us identify and illuminate adaptive problems faced by women. Evolutionary theorists have looked into, among others, the evolutionary basis of patriarchy (see next section), polygyny (Hrdy 1999b; Low 1992, 2000), the relationship between social stratification and the status and treatment of women (Low 1992, 2000), femicide (Daly and Wilson 1988; Wilson, Daly, and Scheib 1997), male sexual proprietariness (Buss 2000; Smuts 1995, 1996; Wilson and Daly 1992), and male sexual coercion (Gottschall and Gotschall 2003; Malamuth 1996; Mesnick 1997; Shields and Shields 1983; Thornhill and Palmer 2000; Thornhill and Thornhill 1983, 1990; Thornhill, Thornhill, and Dizinno 1986).

Their findings can heighten awareness of the risks women face, such as harassment, battering, rape, and exploitation, as well as of the contributing risk factors. Wife abuse, for instance, has been found to be much more frequent in patrilocal societies, where women leave at marriage to live with their husband's kin. The presence of kin apparently protects a woman from male violence (Smuts 1996). The presence of children sired by a former partner has also been identified as a risk factor for violence against wives. This factor was overlooked by traditional researchers, but a Darwinian perspective led to its identification (Daly and Wilson 1996). Likewise, having a stepparent has turned out to be the most powerful predictor of child abuse risk, something that was only discovered after evolutionists started looking (Daly and Wilson 1985, 1988, 1995). Evolutionary findings can also highlight possible sources of patriarchal bias in legal codes, such as in the rape laws of the past when the blame for rape was put on the victim (Thornhill and Palmer 2000).

Evolutionary theory can assist us in specifying what social variables influence individual development in what ways, so that fruitless hypotheses can be identified and selected away. The proposition that the viewing of violent pornography inspires imitative behavior, for example, has no sound theoretical foundation. It is refuted by everything we know about human motivation and development, and it has a logical flaw as well, since it cannot explain why men should want to imitate violent pornography but not the many other human activities depicted in videos, such as having a loving and respectful relationship. Although violent pornography may be a factor in the explanation of raping behavior in some men, it is highly improbable to be a cause of rape in itself (Thornhill and Palmer 2000). This argumentation holds for violence in general: Postulating media violence as a source of mindless imitation is—

besides being no explanation at all, but a labeling—a naive and most doubtful view that is not based on any sound psychological theory (Daly and Wilson 1988).

Evolutionary insights can thus be a valuable aid not only to psychological science, but also to sociology and the political sciences, since they can illuminate how individual actors will be affected by social, economic, and political variables. They can also resolve psychology's puzzlement about the fact that humans are intentional beings, having inbuilt perceptual and cognitive biases, drives, and goals (Daly and Wilson 1996).

An evolutionary framework can elucidate why certain arrangements might not be in women's best interests. It is predictable, for instance, that the 1950s model of the nuclear family, with the male as provider and the female staying home all day looking after the children, should have been frustrating to many women. In contrast to what feminist critic Hilary Rose purports, evolutionary psychology does not at all produce a construction of the family looking "embarrassingly like the Flintstones" (2000:118). As Robert Wright (1995) and Sarah Hrdy (1999b) argue, we can expect that many women will feel the need to actively participate in socioeconomic life. In hunter-gatherer societies women are workers as well as mothers, taking their children with them while gathering, or entrusting them temporarily to the care of relatives. Female primates in general have always been dual-career mothers, forced to compromise between maternal and infant needs, Hrdy writes. For this reason primate mothers, including human foragers, have always shared care of offspring with others.

Evolutionary psychology makes statistical observations and predictions; it says nothing about what people should do nor about what particular individuals will do. If some people feel happy in a Flintstones family, there is nothing wrong with that. Whether a woman wants to stay home with the children, prefers to remain childless, or aspires to both a career and children, she should in each case have that possibility. An evolutionary perspective informs us that acknowledging infant needs does not necessarily enslave mothers: Mothers are important, but they are not the only ones who can provide the caring, nurturing relationship that children need to survive.

Thus feminists need not be afraid that evolutionary insights will dictate any particular way to go for women. In the 1970s a number of feminists dismissed the findings of an infant's need of security and attachment as just another attempt to validate patriarchal interests. They feared that this meant their role as mothers was laid down for years to come. In 1980 the French philosopher Elisabeth Badinter argued that maternal love is merely the product of cultural conditioning (Hrdy 1999b). Such escape routes are not necessary; women do not have to choose between being a mother and making a career. Or at least they should not have to. In practice structural disadvantages in some areas, such as the late-hour meetings in politics, still complicate this choice.

"By political action against femicide and other costs that men impose on women, feminism has made a substantial contribution to reducing the costs of being a woman," Wilson, Daly, and Scheib write. This agenda, they continue, "can be further advanced by scientific understanding of evolved psychological mechanisms, including those masculine sexual and coercive psychological adaptations whose normal and abnormal manifestations conflict with women's evolved psychological adaptations for mate choice and personal autonomy" (1997:458). A feminist agenda need not conflict with knowledge of our evolved human nature. We decide how to apply this knowledge:

> The biological reality is there; its interpretation and implication is what humans may make of it. To deny or ignore biology for fear of it being used against gender, race, and so forth, is to lose the power and freedom of understanding. (Waage 1997:548)

The Evolutionary Origins of Patriarchy

Evolutionary psychology can add a new dimension to the feminist analysis of patriarchy. Many feminists tend to interpret male dominance as a cultural construction and as an end in itself. Evolutionary psychologists, in contrast, assume the roots of patriarchy to lie far back in time, predating the emergence of the human species. Male dominance is, after all, a widespread phenomenon among mammals. From an evolutionary point of view the moving spirit behind it is not male desire for power, but sexual selection. On this account male desire to control females and female sexuality ultimately goes back to the reproductive value of females as the more-investing sex.

It does not follow that patriarchy is inevitable; biological difference does not map directly onto political arrangements. Evolutionists agree that the historical origins of patriarchy cannot be pinned down to one factor, but that it concerns a series of mutually reinforcing evolutionary, social, and cultural developments.[12] This means that to dismantle patriarchy several simultaneous social changes will have to be implemented.

Any reconstruction of how patriarchal arrangements may have unfolded over evolutionary time, as well as of human social evolution in general, remains hypothetical, but we can make some educated guesses. Thus far evolutionary attempts at reconstructing the historical development of patriarchy have focused on the same number of key factors, with male and sometimes female reproductive interests as the underlying evolutionary rationale: patrilocality (when females disperse from their kin at marriage, whereas males stay in their natal community), male coalition-forming, male control over resources, female complicity, and gender ideology. I present the views of three evolutionists in particular: Barbara Smuts, David Buss, and Sarah Hrdy. Although they concur on most accounts, their assessment of the relative importance of some variables varies.

Both Smuts and Hrdy consider our ancestors' presumptive custom of patrilocal residence as the starting point of a series of related practices that allowed males to increase their control over female sexuality and that eventually gave rise to patriarchy. This custom, Smuts (1995) suggests, had its roots in our prehominid forebears.

She reaches this conclusion after examining variations in the patterns of male coercion and female resistance in nonhuman primates. In many primate species, as well as in many other mammals, males use coercive tactics to increase their sexual access to sexually receptive females, such as the threat of violence, aggressive assault, disruption of her copulation attempts with rivals, and infanticide—the latter to bring her back into estrus sooner. Female primates resist this interference with their own sexual goals in a variety of ways. In some species they form coalitions with their female kin to chase off males who threaten or attack a female or her infant. In these female-bonded species male aggression is constrained by the cooperation of females, and by the fact that in these species males need the support of high-ranking females to achieve and maintain a high dominance position. As a result they cannot afford to provoke a high-ranking female's anger. Another way in which females may reduce their vulnerability to male harassment is by forming friendships with one or two particular males. These friends protect her and her infant, and in turn she often cooperates when they try to mate with her.

So the fact that male primates typically are larger than females does not imply that they come off best when conflicts emerge; physical size is no automatic determinant of control. Then what factors mediate the degree of male violence and female resistance? Smuts (1995) finds one variable to be of particular importance: social support from relatives and friends. When this is lacking, males get the upper hand. Two combined factors decide on the presence of this support: matrilocality (when the males, not the females, disperse away from their natal group at adolescence) and female bonding. If the females can stay with their kin, they can form female-female coalitions. If they have to integrate into a new group, on the other hand, bonding will be far more difficult.

What makes Smuts think that patrilocality characterized our hominid forebears? Well, most primates and, in fact, mammals in general, are *matrilocal*, and they are able to thwart the frequent manifestation of male aggression. In contrast to most mammals and primates, however, our closest living relatives—the orangutan, the gorilla, the common chimpanzee, and the bonobo—are patrilocal, apparently because of ecological factors.[13] And also in contrast to most primates, our closest relatives do not seem to be able to prevent or thwart male coercion, with the exception of bonobos. Female gorillas are vulnerable to infanticide when "their" male dies, in orangutans forced copulation appears to be the rule, and chimpanzee females as well are frequently sexually assaulted. Only in bonobos the rate of male aggression against females is extremely low, maybe because they have time for bonding: They can spend most of their time in the company of other females due to a different pattern of food distribution. Bonds between bonobo females are maintained through frequent homosexual activity, and the resulting female-female coalitions against males appear to be so strong that males hardly dare to provoke females (Smuts 1995).

These findings suggest that if there are no relatives around that can offer protection, and if, moreover, females have little opportunity for bonding, the stage is set for male coercion. Most of today's traditional societies show patrilocal residence too, so Smuts's inference is that this practice was probably customary with our hominid ancestors as well. However, as Hrdy (1999a) writes, hunter-gatherers typically have flexible residence patterns. They tend to be patrilocal when intergroup raids are frequent—patrilocality favors male bonding. If they are patrilocal, a system often combined with patrilineal descent, they are typically polygynous and more coercive concerning female sexuality. Hrdy believes it must have been agriculture that gave rise to patrilocal residence patterns. Whatever the time span, both Smuts and Hrdy agree that in the course of history the rise of patrilocal residence, in addition to the fact that for reasons unknown human females do not typically form strong coalitions, was the first step in the disempowerment of females.[14]

As a result of patrilocality, males were able to form long-term alliances with each other, presumably to increase their chances of success in competing against males from other groups and to protect defensible resources, whether women or, once agriculture was established, land and eventually cities. In addition to protecting females from abduction by males from other communities, these alliances could also be used to control them. Intensive agriculture must have increased male power over women, since the latter lost their subsistence role and independency as gatherers. With their movements confined to a small piece of land, they were easier to coerce (Smuts 1995).

Indeed, a case study of the !Kung San in the Kalahari desert, presented in Wilson (1978), provides a dramatic illustration of this. This hunter-gatherer community is characterized by egalitarian sex roles. Women gather while men hunt (although the tasks are often shared), and both sex roles are respected by all. Girls and boys are apparently not differentially socialized, but small average differences still appear. A few !Kung San bands, however, have settled into villages and taken up farming. The work is heavier, and, unlike in the hunter-gatherer groups, younger children are expected to significantly partake in it. The girls stay at home to look after smaller children and to do the household work, whereas the boys take care of the gardens and the cattle. By maturity there are marked sex differences in lifestyle and status: Women are domestic and are continuously supervised, whereas men can freely wander around. Women's status has plummeted considerably.

When resources became accruable it became important for men to try to ascertain that the children in which they were investing were really their own, Smuts (1995) continues. Male support of offspring in turn made it costly for women to stand up against male domination.

The agricultural revolution in combination with increasingly well-developed male political alliances resulted in inequalities in male wealth and power. Powerful males became more able to impose their will upon females, since they also dominated other males, who now no longer had the power to intervene. Males at the top could increasingly use their power to monopolize females (eventually leading to harems) and so reduce the mating opportunities of other males. According to Smuts, this hypothesis fits in with the observation that cross-culturally the extent of male domination of women is positively associated with the degree of hierarchy among men. Female complicity must have been another critical factor in the evolution of patriarchy: Often females could better pursue their material and reproductive interests by competing with other females and allying with males than by allying with other females. A final crucial factor must have been the evolution of language, which led to the development of gender ideologies (Smuts 1995).

Female complicity in patriarchy is accepted as a fact by all three theorists. They defend different points of view, however, when trying to identify the causal forces producing this complicity. According to Smuts (1995), male coercive behavior eventually led to male resource control, with women more or less being forced to participate in this system if they wanted to get something out of it for themselves and their children. Buss's (1996) analysis focuses on male resource control as largely a product of the coevolution of women's mate preferences and men's competitive strategies. In this view women's preferences for a mate with status and resources established the ground rules for men in their competition with one another. Men started controlling resources in order to attract women, and women's preferences followed, in turn selecting for more ambition, competitiveness, and risk-taking in men. The intertwining of these psychological mechanisms in women and men created the conditions for male domination in the domain of resources (Buss 1996).

Buss's theory receives support from the finding that, as mentioned in the section on female psychology, in hunter-gatherer societies, where no resources are accrued, high-status males have most success in attracting many females, which suggests psychological adaptations in women that generate sexual attraction to traits associated with social status, and from the finding that women worldwide continue to express a preference for marrying men with more resources. Hrdy (1997, 1999a, 1999b) objects to this interpretation. She agrees with Marxist theorists that male control over productive resources that women need to reproduce was forced upon women, and that this control led to the development of full-fledged patriarchy. It was not female choice that led to male resource accumulation, she argues, since among primates, male control of resources (territories) *preceded* female choice for males with resources. Males just happen to control territories due to differential male and female migration patterns, patrilocality, male-male coalitions, and greater male strength. In such a system we may expect females to choose mates on the basis of the resources they control, but this does not mean that female choice *was responsible* for male competition or male control of resources. Because male competition and male resource control were already present in the hominid line, we may almost be certain that they preceded the development of female preferences for resource-holding males in humans, she concludes. Women's continuing preference for men with resources is a facultative response to their more unsure economic position than men's (Hrdy 1997).

The problem with this view is that women who themselves have access to resources do not tend to value status and resources in a mate less, but *more*, as we have seen. Nevertheless, there is no need to see Hrdy's and Buss's hypotheses as mutually exclusive. Hrdy presents it that way, though: "Choices that evolutionary psychologists studying mate choices interpret as natural outcomes of women's own innate preferences are to others, including me, the outcome of 'patriarchal constraints' on female reproductive options" (1999a:xxvi). Buss (1996), however, writes that evolutionary psychologists expect context sensitivity of behavior and that, therefore, if women were to shift their mate preferences when they gain full economic equality, this would not be incompatible with evolutionary models. It would, however, require a more complex depiction of our evolved desires.

To conclude, the evolutionary perspective holds that patriarchy ultimately revolves around the control of female sexuality, and it explains why this is so: because women are the more-investing, limiting sex. Thus the radical feminist observation that sexual control lies at the heart of patriarchy (Brownmiller 1975; Dworkin 1997; French 1992) is basically correct. Men are, however, not likely to be united in their interests. They want to control reproductively valuable women, but they also compete among each other for women and for status and resources, because these assets help them to attract the other sex. Men form coalitions, but they do this primarily in order to out-compete other men. Women are not united in their interests with members of their own sex either. They compete among each other for high-status males. Their weapons are the evolved mate preferences of men, such as youth and beauty (Buss 1996; Buss and Duntley 1999).

This does not mean that women are to blame for the rise of patriarchy. As Buss (1996) explains, the notion of blame is senseless when describing a process that unfolded over the course of evolutionary time. It does imply, however, that the forces producing patriarchy are still at work in the present. Both sexes contribute to its perpetuation, not only men. Women do so by, for instance, continuing to want status and resources in a male, by competing with other women or allying with males (Campbell 2002; Hrdy 1999b), by sometimes socializing their sons and daughters in ways that perpetuate male dominance (Low 1989), and by guarding their domestic power and displaying possessiveness toward their children (Lopreato and Crippen 1999).

This complex scheme of things does not undermine the legitimacy of many feminist claims, but it reminds us that women are not puppets on a string; they have their own sources of power and influence. If women want to fight patriarchy successfully, they will have to renounce the behaviors that contribute to its perpetuation. Following Smuts's analysis of patriarchy, other directions to take are the further development of female political solidarity and of laws and institutions that protect women from male coercion, the creation of economic opportunities for women, female support of policies that reduce inequality among men, and feminist consciousness-raising. As she acknowledges, these measures are of course already part of the feminist agenda. This shows again that feminist and evolutionary perspectives can be integrated. An evolutionary perspective can strengthen feminism by identifying the ultimate causes of the male desire to control women. Furthermore it helps to guide analysis of systems of gender inequality, by suggesting that we need to ask "how it affects female sexuality and reproduction in ways that benefit some men at the expense of women (and other men)" (Smuts 1995:22). With the evolutionary analysis of patriarchy still being in its infancy, it can use the help of feminist theorists in order to mature.

Toward a Darwinian Left

Modern Darwinian theory stresses cooperation as well as competition. Evolution may well have endowed us with selfish and greedy tendencies, but we have also evolved the capacity, even the need, to love, respect, and help others. Most of us empathize automatically when we see others in pain, even if they are only fictional characters in a book or television series. The fact that the ultimate roots of emotions like compassion and altruism are that they helped our ancestors to survive and reproduce, does not detract from the reality and authenticity of these feelings.

This is part of the reason why evolution can be invoked to argue for policies ranging from the right to the left: It just depends which evolved propensities you prefer to focus on in establishing your view of human nature. The other part is that the kind of society you choose to strive for will depend on your set of values: whether you value care over competition and communality over individual freedom, for instance. Darwinian theory cannot guide us here, since it is a scientific theory that tries to explain how things came to be. It does not advise us on what we should do. Although in the past it has been appropriated foremost by the political right, it has by itself no obvious political implications.

In *A Darwinian Left* (1999) philosopher Peter Singer argues for a political left that does not deny the existence of a human nature and that has replaced its utopian ideas by a more realistic view of what can be achieved. A Darwinian left would want policies to be grounded on the best available evidence of what human beings are like. It would acknowledge that "an understanding of human nature in the light of evolutionary theory can help us to identify the means by which we may achieve some of our social and political goals, including various ideas of equality, as well as assessing the possible costs and benefits of doing so" (1999:15). Singer uses the metaphor of wood carvers having to make a bowl: Instead of simply beginning to carve, they will let their design be informed by the specific wood they are to work with in order to suit its grain. Political philosophers and revolutionaries, however, have typically designed their ideal society without taking into account the human beings who had to live with and carry out their plans. A better way, Singer says, would be to tailor your policy to the tendencies inherent in humans in order to get the best out of them.

This means that we should give up the dream of the perfectibility of humankind. Humans are malleable, but this malleability has its limits. Steven Pinker (2002) sums up some scientific findings that make the emergence of a utopian society highly unlikely. Among them are the primacy of family ties in all human societies, with their consequent appeal of nepotism and inheritance; the universality of dominance hierarchies and violence in all human societies, even in hunter-gatherer groups; the universal preference for kin and friends; the universality of group-against-group hostility, and the small effort it takes to arouse it; the partial heritability of many personality traits, including intelligence, conscientiousness, and antisocial inclinations, implying that even perfectly equal treatment will result in unequal outcomes; and the emergence of social loafing when chances are that it will go undetected.

A Darwinian left would take these discoveries into account. It would stand by traditional leftist values by being on the side of the weak, the poor, and the oppressed, and by prioritizing equality as a moral or political ideal, but, as Singer (1999) explains, it would think very carefully about what social and economic changes will really implement these goals. It would not expect to end all conflict between human beings, but it would believe that most people will react positively to genuine opportunities to cooperate in mutually beneficial relationships. It would thus build on the social and cooperative side of our nature without ignoring the individualistic and competitive side. It would also bear in mind the evolved psychology of the sexes. It would reject any inference from what is "natural" to what is "right," and work toward a less-anthropocentric view of animals and the whole of nature (Singer 1999).

Singer's outline of this new left is sure to be found objectionable by many traditional leftists. Whereas the latter assume that all inequalities are due to discrimination, prejudice, oppression, or social conditioning, Darwinian leftists would not automatically do so. They would recognize, for instance, that the underrepresentation of women in the upper regions of the political and economic hierarchies may partially result from men's greater eagerness and opportunism in pursuing social status and political power. Men are much more ready than women to sacrifice their family and social relationships for the sake of getting to the top (Browne 1998). This does not imply that Darwinian leftists would defend the status quo, since it can safely be assumed that factors such as gender discrimination are still in play as well.

In an analysis of peer-reviewed scores for Swedish postdoctoral fellowship applications, for instance, Christine Wennerås and Agnes Wold (1997) revealed the academic world as being highly prejudiced against women. The fact that these results were found in Sweden, a leading country in the world with respect to equal opportunities for women and men, only makes the results of this study the more alarming. Wennerås and Wold found that female applicants had to produce two and a half times more scientific papers than male applicants to be conceived as equally competent. Lacking personal affiliations with one of the reviewers had an effect of the same magnitude: Applicants without these ties had to be two and a half times more productive than applicants having them. Personal affiliations were as frequent for male as for female applicants, but this means that a woman without connections had to be five times more productive than a man affiliated with one of the committee members.

Several other studies have shown that an academic paper said to be written by a man is rated more positively by both women and men than when attributed to a female author (Schiebinger 1999). Experiments in which subjects had to hire (fictitious) applicants for hierarchy-enhancing and hierarchy-attenuating jobs revealed that they partially relied on sex stereotypes to decide what kinds of jobs women and men would do well (Pratto 1996). These are just a few examples to indicate the thriving of gender bias. A Darwinian left will have a difficult task in disentangling the many complex factors affecting the differential distribution of women and men in society.

Another delicate question to be addressed is what we mean exactly when we talk about "equality as a moral or political goal." As Janet Radcliffe Richards (1997) points out, the very fuzziness and elusiveness of the concept of equality may be the reason why so many people are in defense of it. People mean different things by the term, or they slip from one definition of equality into another without realizing it. Some think, for instance, that equality of opportunity automatically implies equality of outcome. As a result the debate about equality of opportunity gets very confused.

To some the term just implies impartiality of treatment. To others it means something qualitatively different: the right to move and rise in society according to one's abilities and inclinations. Whereas in the first definition nothing needs to be done except for the removal of arbitrary preferences, the second definition demands that you specify which policies you want to pursue to the detriment of other policies. You cannot equalize *all* opportunities over *all* people. Even in the simple example of equal opportunity for all children in a classroom, a teacher will have to choose between, for instance, making herself equally available to all, giving equal time to all, giving more to children of deprived backgrounds, or trying to make all children equally proficient in everything. In the last two cases she will have to treat individuals unequally in the name of equality.

Some have yet another interpretation of equality of opportunity. To them it means the equalizing of backgrounds, so that everybody has the same starting conditions. This definition moves us to yet another standard of opportunity. We slip from ideas about equality of opportunity as *something specific* that people *give*, such as admission to a school or admission to applicate for a job, to opportunity as *something general and undirected* that people *have*, in the way of the ability to make their lives take whatever direction they want it to take. The way opportunity is interpreted here is thus completely different. It becomes a rough term for the ability to arrange your life whatever way you want it to, because of the good starting conditions that you find yourself in, rather than for the ability to achieve an end predetermined by others, such as getting to the top of the political hierarchy as a woman. Most people confuse these meanings. If we try to clarify exactly what we mean, we might see that the problem changes, Radcliffe Richards argues. If we realize that actually we regard "opportunity in an undirected sense," that is, the opportunity to live your life the way you want it to, as a good in itself, the relevant question just becomes how to increase its overall amount. This means, however, that the concept of equality will become less important. If we have a fair distribution of "opportunity to live your life the way you want to," we cannot expect results to be equal, because people are different. Even if we equalize *all* starting conditions, we still would get differential outcomes, since people have different genetic endowments (Radcliffe Richards 1997).[15]

These are powerful thoughts. It is primordial for the left to consider analyses like these, and even more crucial for a Darwinian left, since for Darwinians the subject is complicated by their recognition of the partial heritability of many personality characteristics and abilities, as well as of average differences between the sexes.

Notes

1. The practice of secluding women from the sight of men except for the head of the family, by curtains in the home or by veils outside it.

2. Madeline Rich, director of the YWCA Chicago, wrote in her review of *A Natural History of Rape* (Thornhill and Palmer 2000): "Rape is no more about sex than hitting someone over the head with a frying pan is about cooking." Many misunderstandings arise, however, from the multiple possible meanings of each word in the phrase "Rape is about violence, not sex," as legal scholar Owen Jones (2000) explains. Does the phrase refer to violence as the cause of rape, or as its meaning to the victim, for instance? Does it mean that rape is not at all about sex, or not primarily, or not on average? And so on. The ambiguity of the question "Is rape sex or violence?" is also laid out by the feminist researcher Charlene Muehlenhard and her colleagues (1996).

3. Peak periods with testosterone levels being much higher in males than in females are found in humans as well as in other mammals. In humans, testosterone peaks again in males in the period from right after birth until about five months of age, and at the time of puberty.

4. Although most physical sex differences in humans seem to result from sexual selection, some physical dimorphisms, such as women's wider hips, have likely evolved through natural selection. Neonates' bigger brains required a wider pelvic region to pass through. Once this characteristic evolved, however, males apparently started to consider it sexually attractive. The result is that the size of women's hips began to influence mate choice and is thus affected by sexual selection as well. Another dimorphism that might be largely due to natural selection is women's finer manual dexterity, which might be related to the hunter-gatherer lifestyle of our ancestors, with women doing most of the gathering (Geary 1998).

5. Readers who want to delve more deeply into the evolved differences between the sexes are recommended one of the following exciting books: *The Evolution of Desire* (David Buss, 1994), *The Science of Romance* (Nigel Barber, 2002), *The Essential Difference* (Simon Baron-Cohen, 2003), or, in a more academic vein, *Male, Female* (David Geary, 1998), and *Sex Differences* (Linda Mealey, 2000).

6. Hrdy (1999b) objects that in other primates no such male preference for youthful features in females is found. Primate males generally prefer older females, because these are more experienced mothers and therefore have more surviving offspring. She suggests that in humans this male preference for youth and beauty may be due to patriarchy. Primate females, however, do not experience menopause; women do. We might therefore expect men to be much more sensitive to age than other male primates. Furthermore, since humans are a species forming long-term romantic bonds, men should prefer mates that have the highest fecundity (moderated by issues of compatibility) (Geary 1998).

7. *Near*-universal, because other variables may influence the nature of male preferences, such as population-typical variations in the size of WHR, or harsh ecological conditions, which may make men prefer women with a greater amount of fat to women with low WHR (Singh 2002).

8. Youth and beauty are central features of homosexual men's mate preferences as well. This suggests that not even variations in sexual orientation alter these deeply rooted psychological mechanisms (Buss 1994).

9. Homosexual men tend to have a much higher promiscuity level than heterosexual men, probably because they are not thwarted in fulfilling their evolved desires by the differential sexual psychology of the opposite sex, as first suggested by Donald Symons in *The Evolution of Human Sexuality* (1979) (Bailey 2003).

10. Synopsis on back cover of *Seduction* by Jayne Ann Krentz.

11. The ratio of sexually active males to sexually active females in any given population at a given point in time. When the sex ratio is female biased, that is, when there are more sexually active women than men, liberal sexual attitudes tend to prevail, presumably because then men are more in a position to impose their wishes upon women, since they are in scarcer supply than women (Schmitt et al. 2003a).

12. Buss 1994, 1996, 1999; Buss and Duntley 1999; Fisher 1999; Gowaty 1992; Hrdy 1997, 1999a, 1999b; Pratto 1996; Smuts 1995; Wilson 1978.

13. The same holds for hamadryas baboons, who are very vulnerable to infanticide by males (Smuts 1995; Hrdy 1999a).

14. Matrilocal and matrilineal societies, in which women have much more freedom, are typically hunter-gatherer or horticultural societies, not societies that know animal husbandry or that are acquainted with the plow (Hrdy 1999b).

15. For a clear-cut analysis of the multiple ways in which the word "equal" is used in the political debate, see Radcliffe Richards (1998).

Conclusion

As I hope to have shown, condensed and generalized though my outline was, feminism and evolutionary psychology need each other. The history of evolutionary science attests to its liability to all kinds of bias, as does the history of the social and life sciences in general. Only the input of women researchers (and of people from differing ethnic, class, and sexual backgrounds) can guarantee our knowledge of human nature to become as value-free as possible.

Feminism, however, is missing a unifying framework. What we have are multiple and often mutually exclusive accounts of human nature, as well as endless discussions on the problems of "essentialism" and of equality "versus" difference, problems that turn out to be pseudoproblems when considered from a biologically informed perspective. As feminist philosopher Elizabeth Grosz points out in her exploration of a possible alliance between Darwinism and feminism, "Darwin's work offers a subtle and complex critique of both essentialism and teleology. It provides a dynamic and open-ended understanding of the intermingling of history and biology . . . and a complex account of the movements of difference, bifurcation, and becoming that characterise all forms of life" (1999:33–34). Elizabeth Wilson is another one of those rare feminists calling for an evolutionarily inspired feminism, arguing that the biological sciences can actually sustain feminist claims and that feminists should be open-minded enough to try to *learn* from the data produced rather than just wanting to uncover sources of bias (1999, 2002).

Indeed, the logical inconsistencies and the lack of openness toward or ignorance of other than environmentalist accounts of gender difference that typify many feminist theories can harm the intellectual credibility of the movement. If feminists continue to reject the mounting evidence from the biological sciences (e.g., genetics, behavioral genetics, neurophysiology, endocrinology, evolutionary biology) and the social sciences (e.g., cognitive psychology, neurolinguistics, artificial intelligence) with regard to the biological underpinnings of human behavior, they back themselves into an embarrassingly uninformed corner. In order to understand human nature and human behavior, *all* possible sources of information should be taken into account, not just those that exert an ideological appeal to us.

"History has not been kind to ideologies that rested on patently false beliefs about human nature," science writer Robert Wright notes (1994a:34). He is referring to communism and Lysenkoism, but also to the notion of gender as a mere social construct. The tone of his article is rather chiding. On reading it carefully, however, it becomes clear that Wright is no antifeminist. He even looks for rationales for affirmative action based on our emerging knowledge of evolved sex differences. Thus, following a suggestion by Matt Ridley, he proposes that we might favor quotas for women because they are less ready than men to sacrifice an organization's welfare to personal advancement. Women's political priorities are also known to differ from men's: Women are more pacifistic and more concerned about reducing social inequalities (Fisher 1999), a finding that might be used as a rationale for affirmative action in politics. Many possibilities of applying evolutionary insights in ways that benefit women are open to exploration. An example is the argument proposed by Alexander Sanger, grandson of social reformer and founder of the American birth control movement Margaret Sanger, in his book *Beyond Choice* (2004). Sanger develops an evolutionary defense of reproductive freedom, showing that the use of methods to prevent pregnancy and childbirth is no modern aberration, but is profoundly human. Women have always done what was necessary to have their children survive, and one of their means to that end has always been the timing and spacing of childbearing. An evolutionary perspective, according to Sanger, supports both reproductive freedom and respect for human life. Women and men have been successfully controlling their reproduction for a long time. To oppose the legalization of contraception and abortion is simply to "show an abysmal ignorance of biology and human nature" (Sanger 2004:138). It might be a point of debate, however, whether Sanger is not committing the naturalistic fallacy.

Nobody contends that women are saints. They just typically score higher than men on qualities such as nurturance and communality, whereas men typically outstrip women when it comes to qualities like agency and systemizing. These differences are not likely to stem merely from culture and socialization. Why on earth should we want to deny or eradicate them, instead of acknowledging the difference and getting the best of it? We are no longer living in a pre-Darwinian world with its essentialist assumptions about a 'natural order' of the sexes, fixed in rigid structures such as the so-called Great Chain of Being. Science has shown that women and men are more similar than they are different, thus shattering many old prejudices about "the sex," as women were called in the nineteenth century. This finding, however, should not lead us to ignore the differences that exist. All people have more in common than they are different, yet as far as I know nobody is standing on the barricades to demand that all people should be considered identical.

Wright hits the mark in saying that, if feminists—rightfully—want laws that protect women against sexual harassment, they will have to admit that women are in some ways uniquely vulnerable. Men will not feel violated as soon and as deeply by unwanted sexual advances by the other sex, and for good evolutionary reasons. If you want to give women more protection than men, you will therefore have to acknowledge the specificities of the female mind (Wright 1994a). Now *this* is a truly feminist remark, I would say: In defense of women in the domain where the 'battle of the sexes' is fought most fiercely: that of sex.

Indeed, the observation that somebody rejects an environmentalist account of gender difference does not imply that he or she does not sympathize with the feminist cause. On the contrary, it is precisely as a feminist that one might worry most about the antiscientific tendencies and the biophobia within feminism. A lot of valuable energy that might be directed at fighting the real problems faced by women is now wasted on elaborate, far-fetched, and sometimes unintelligible accounts of the 'social construction' of scientific knowledge and of gender identity.

It is true, though, that evolutionary insights might be misused by political conservatives. That makes it all the more important for feminists to enlighten the public about the misconceptions that easily come with Darwinian theory, such as the naturalistic fallacy. The caricatures of evolutionary psychology found in feminist writings only reinforce these misconceptions, however, thus paradoxically enhancing their potential appeal to the political right. The resistance against biologically informed explanations of the human mind and human behavior indeed carries the implicit message that if these contested theories turn out to be irrefutable, then it would be okay to discriminate against women (or gays, or . . .) after all. This, of course, is absurd. Biological facts do not, by themselves, dictate our values.

I think that the time has come for feminists to shed their unreasonable fears of a Darwinian framework. We no longer live in the nineteenth century, when male scientists could develop unsubstantiated theories about female nature without even listening to women's experiences. We should not let ourselves be held hostage to the past. Science has grown up. At the same time it remains a human and therefore fallible enterprise. Keeping an eye open to gender bias in evolutionary theories will continue to be required, just as awareness of other possible sources of bias will always be needed. This should not keep us from acknowledging that the paradigm of evolutionary psychology is basically sound, although, as a young scientific discipline, it may still show some growing pains. We will have to take its data and the data from other scientific disciplines into account if we want our gender politics to be attuned to what women and men find important in their relationships and in their lives.

I could not put it more aptly than Anne Campbell does in *A Mind of Her Own*:

> If . . . we accept that women and men are different, we can think about a society that breaks down the barriers between children and work, that allows women to see value in cooperation as well as competition and that allows women to capitalize on their linguistic advantages. If evolutionary theory is correct then we cannot design twenty-first century woman as if from scratch. Ideology, social policies, law and the media cannot in and of themselves make women into something they are not. What we can and should do is to give people choices that allow them the maximum freedom to be whatever they want. With that freedom, women's nature can take its own course. (2002:32–33)

How we should do that still has to be worked out in detail. A major task is awaiting feminism.

Bibliography

Abbey, Antonia, et al. 1996. Alcohol, Misperception, and Sexual Assault: How and Why Are They Linked? In *Sex, Power, Conflict: Evolutionary and Feminist Perspectives*, eds. David Buss and Neil Malamuth. New York: Oxford University Press.

Alexander, Michele, and Terri Fisher. 2003. Truth and Consequences: Using the Bogus Pipeline to Examine Sex Differences in Self-Reported Sexuality. *The Journal of Sex Research* 40(1):27–35.

Allen, Caitilyn. 1997. Inextricably Entwined: Politics, Biology, and Gender-Dimorphic Behavior. In *Feminism and Evolutionary Biology: Boundaries, Intersections, and Frontiers*, ed. Patricia Gowaty. New York: Chapman & Hall.

Angier, Natalie. 1999. *Woman: An Intimate Geography*. London: Virago.

Archer, John. 1996. Sex Differences in Social Behavior: Are the Social Role and Evolutionary Explanations Compatible? *American Psychologist* 51(9):909–917.

Bailey, Michael. 2003. *The Man Who Would Be Queen: The Science of Gender-Bending and Transsexualism*. Washington: Joseph Henry Press.

Barash, David. 1979. *Sociobiology: The Whispering Within*. New York: Harper & Row.

———. 1981. *The Whisperings Within: Evolution and the Origin of Human Nature*. New York: Penguin Books.

Barber, Nigel. 2002. *The Science of Romance: Secrets of the Sexual Brain*. New York: Prometheus Books.

Barkow, Jerome, Leda Cosmides, and John Tooby, eds. 1992. *The Adapted Mind: Evolutionary Psychology and the Generation of Culture*. New York: Oxford University Press.

Baron-Cohen, Simon. 1995. *Mindblindness: An Essay on Autism and Theory of Mind*. Cambridge: MIT Press, 1997.

———. 2003. *The Essential Difference: Men, Women, and the Extreme Male Brain*. London: Allen Lane.

Beldecos, Athena, et al. (aka The Biology and Gender Study Group). 1988. The Importance of Feminist Critique for Contemporary Cell Biology. In *Feminism and Science*, ed. Nancy Tuana. Bloomington: Indiana University Press.

Benedict, Ruth. 1934. *Patterns of Culture*. Boston: Houghton Mifflin.

Benenson, Joyce. 2002. Similarities between Sex Differences in Social Organization in Chimpanzees and Human Children. Unpublished paper presented at Human Behavior and Evolution Society Conference, New Brunswick, N.J., June 19–23.

Benton, Ted. 2000. Social Causes and Natural Relations. In *Alas, Poor Darwin: Arguments against Evolutionary Psychology*. London: Jonathan Cape.

Birke, Lynda. 1999. *Feminism and the Biological Body*. New Brunswick: Rutgers University Press, 2000.

Birkhead, Tim. 2000. *Promiscuity: An Evolutionary History of Sperm Competition and Sexual Conflict*. London: Faber & Faber.

Bjerke, Tore. 1992. Sex Differences and Aggression: The Role of Differential Socialization and Sex Stereotypes. In *The Nature of the Sexes: The Sociobiology of Sex Differences and the "Battle of the Sexes,"* ed. Johan van der Dennen. Groningen: Origin Press.

Blackman, Julie. 1985. The Language of Sexual Violence: More Than a Matter of Semantics. In *Violence against Women: A Critique of the Sociobiology of Rape*, eds. Suzanne Sunday and Ethel Tobach. New York: Gordian Press.

Blackwell, Antoinette. 1875.*The Sexes throughout Nature*. Westport, Conn.: Hyperion Press, 1976.

Bleier, Ruth. 1985. Biology and Women's Policy: A View from the Biological Sciences. In *Women, Biology and Public Policy*, ed. Virginia Sapiro. Beverly Hills, Calif.: Sage Publications.

Bordo, Susan. 1986. The Cartesian Masculinization of Thought. In *Sex and Scientific Inquiry*, eds. Sandra Harding and Jean O'Barr. Chicago: University of Chicago Press, 1987.

———. 1990. Reading the Slender Body. In *Visual Culture Reader*, ed. Nicholas Mirzoeff. London: Routledge, 1998.

Bowler, Peter. 1990. *Charles Darwin: The Man and His Influence*. Cambridge: Cambridge University Press.

Bradley, Susan, et al. 1998. Experiment of Nurture: Ablatio Penis at 2 Months, Sex Reassignment at 7 Months, and a Psychosexual Follow-up in Young Adulthood. *Pedriatrics* 102(1):e9.

Braeckman, Johan. 2001. *Darwins moordbekentenis*. Amsterdam: Nieuwezijds.

Braeckman, Johan, Tom Speelman, and Griet Vandermassen. 2001. Evolutiepsychologie: basisprincipes en misverstanden. *Mores* 46(4):311–320.

Brouns, Margo. 1995a. Feminisme en wetenschap. In *Vrouwenstudies in de jaren negentig. Een kennismaking vanuit verschillende disciplines*, eds. Margo Brouns et al. Bussum: Coutinho.

———. 1995b. Kernconcepten en debatten. In *Vrouwenstudies in de jaren negentig. Een kennismaking vanuit verschillende disciplines*, eds. Margo Brouns et al. Bussum: Coutinho.

———. 1995c. Theoretische kaders. In *Vrouwenstudies in de jaren negentig. Een kennismaking vanuit verschillende disciplines*, eds. Margo Brouns et al. Bussum: Coutinho.

Brouns, Margo, Mieke Verloo, and Marianne Grünell, eds. 1995. *Vrouwenstudies in de jaren negentig. Een kennismaking vanuit verschillende disciplines.* Bussum: Coutinho.

Brown, Donald. 1991. *Human Universals.* New York: McGraw-Hill.

Browne, Janet. 2002. *Charles Darwin: The Power of Place.* Princeton: Princeton University Press.

Browne, Kingsley. 1998. *Divided Labour: An Evolutionary View of Women at Work.* London: Weidenfeld & Nicolson.

Brownmiller, Susan. 1975. *Against Our Will: Men, Women and Rape.* New York: Ballantine Books.

———. 1984. *Femininity.* New York: Linden Press/Simon & Schuster.

Bryson, Valerie. 1999. *Feminist Debates: Issues of Theory and Political Practice.* London: Macmillan.

Bunge, Mario. 1996. In Praise of Intolerance to Charlatanism in Academia. In *The Flight from Science and Reason*, eds. Paul Gross, Norman Levitt, and Martin Lewis. New York: The New York Academy of Sciences.

Burley, Nancy, and Richard Symanski. 1981. Women Without: An Evolutionary and Cross-Cultural Perspective on Prostitution. In *The Immoral Landscape: Female Prostitution in Western Societies*, ed. Richard Symanski. Toronto/Ontario: Butterworths.

Buss, David. 1989. Sex Differences in Human Mate Preferences: Evolutionary Hypotheses Tested in 37 Cultures. In *Human Nature: A Critical Reader*, ed. Laura Betzig. New York: Oxford University Press.

———. 1992. Mate Preference Mechanisms: Consequences for Partner Choice and Intrasexual Competition. In *The Adapted Mind: Evolutionary Psychology and the Generation of Culture*, eds. Jerome Barkow, Leda Cosmides, and John Tooby. New York: Oxford University Press.

———. 1994. *The Evolution of Desire: Strategies of Human Mating.* New York: Basic Books.

———. 1996. Sexual Conflict: Evolutionary Insights into Feminism and the "Battle of the Sexes." In *Sex, Power, Conflict: Evolutionary and Feminist Perspectives*, eds. David Buss and Neil Malamuth. New York: Oxford University Press.

———. 1999. *Evolutionary Psychology: The New Science of the Mind.* Needham Heights: Allyn & Bacon.

———. 2000. *The Dangerous Passion: Why Jealousy Is As Necessary As Love and Sex.* London: Bloomsbury Publishing Plc.

Buss, David, and Joshua Duntley. 1999. The Evolutionary Psychology of Patriarchy: Women Are Not Passive Pawns in Men's Game. *Behavioural and Brain Sciences* 22(2):219–220.

Buss, David, and Neil Malamuth, eds. 1996. *Sex, Power, Conflict: Evolutionary and Feminist Perspectives.* New York: Oxford University Press.

Buss, David, and David Schmitt. 1993. Sexual Strategies Theory: An Evolutionary Perspective on Human Mating. *Psychological Review* 100(2):204–232.

Butler, Judith. 1990. *Gender Trouble: Feminism and the Subversion of Identity.* New York/London: Routledge.

———. 1997. *The Psychic Life of Power: Theories in Subjection.* Stanford: Stanford University Press.

Callen, Anthea. 1998. Ideal Masculinities: An Anatomy of Power. In *Visual Culture Reader*, ed. Nicholas Mirzoeff. London and New York: Routledge.

Campbell, Anne. 2002. *A Mind of her Own: The Evolutionary Psychology of Women.* New York: Oxford University Press.

Chodorow, Nancy. 1978. *The Reproduction of Mothering: Psychoanalysis and the Sociology of Gender.* Berkeley: University of California Press. 2nd ed. published in 1999.

Chomsky, Noam. 1959. A Review of B. F. Skinner's *Verbal Behavior. Language* 35(1):26–58.

Cohen, Mark, and Sharon Bennett. 1993. Skeletal Evidence for Sex Roles and Gender Hierarchies in Prehistory. In *Reader in Gender Archeology*, eds. Kelley Hays-Gilpin and David Whitley. London: Routledge.

Colapinto, John. 2000. *As Nature Made Him: The Boy Who Was Raised as a Girl.* New York: HarperCollins Publishers.

Cole, Stephen. 1996. Voodoo Sociology: Recent Developments in the Sociology of Science. In *The Flight from Science and Reason*, eds. Paul Gross, Norman Levitt, and Martin Lewis. New York: The New York Academy of Sciences.

Connell, Bob. 1995. Gender as a Structure of Social Practice. In *Space, Gender, Knowledge: Feminist Readings*, eds. Linda McDowell and Joanne Sharp. London: Arnold.

Cosmides, Leda, John Tooby, and Jerome Barkow. 1992. Introduction: Evolutionary Psychology and Conceptual Integration. In *The Adapted Mind: Evolutionary Psychology and the Generation of Culture*, eds. Jerome Barkow, Leda Cosmides, and John Tooby. New York: Oxford University Press.

Coyne, Jerry. 2000. Of Vice and Men: The Fairy Tales of Evolutionary Psychology. *The New Republic*, April 3.

Crews, Frederick. 1996. Freudian Suspicion versus Suspicion of Freud. In *The Flight from Science and Reason*, eds. Paul Gross, Norman Levitt, and Martin Lewis. New York: The New York Academy of Sciences.

———, ed. 1998. *Unauthorized Freud: Doubters Confront a Legend.* New York: Penguin Books.

Cronin, Helena. 1991. *The Ant and the Peacock: Altruism and Sexual Selection from Darwin to Today.* New York: Cambridge University Press, 1994.

———. 1997. It's Only Natural. *Red Pepper*, July.

Cunningham, Michael, et al. 1995. "Their Ideas of Beauty Are, on the Whole, the Same as Ours": Consistency and Variability in the Cross-Cultural Perception of Female Physical Attractiveness. *Journal of Personality and Social Psychology* 68(2):261–279.

Curry, Oliver. 2002. A Cut-Out-and-Keep Guide to the Naturalistic Fallacy. Unpublished paper presented at HBES conference, New Brunswick, June 19–23.

Daly, Martin, and Margo Wilson. 1985. Child Abuse and Other Risks of Not Living with Both Parents. *Ethology and Sociobiology* 6:197–210.

———. 1988. *Homicide.* New York: Aldine de Gruyter.

———. 1995. Discriminative Parental Solicitude and the Relevance of Evolutionary Models to the Analysis of Motivational Systems. In *The Cognitive Sciences*, ed. Michael Gazzaniga. Cambridge: MIT Press.

———. 1996. Evolutionary Psychology and Marital Conflict. In *Sex, Power, Conflict: Evolutionary and Feminist Perspectives*, eds. David Buss and Neil Malamuth. New York: Oxford University Press.

Darwin, Charles. 1964 [1859]. *On the Origin of Species.* Cambridge: Harvard University Press.

———. 1998 [1871/1874]. *The Descent of Man; and Selection in Relation to Sex.* 2nd ed. New York: Prometheus Books.

———. 1887. *The Autobiography of Charles Darwin, 1809–1882*, ed. Nora Barlow. New York: Norton.

Dawkins, Richard. 1976. *The Selfish Gene.* Oxford: Oxford University Press.

de Beauvoir, Simone. 1949. *The Second Sex.* Trans. H. M. Parshley. New York: Penguin Books.

Denfeld, Rene. 1996. Old Messages: Ecofeminism and the Alienation of Young People from Environmental Activism. In *The Flight from Science and Reason*, eds. Paul Gross, Norman Levitt, and Martin Lewis. New York: The New York Academy of Sciences.

Denmark, Florence, and Susan Friedman. 1985. Social Psychological Aspects of Rape. In *Violence against Women: A Critique of the Sociobiology of Rape*, eds. Suzanne Sunday and Ethel Tobach. New York: Gordian Press.

Dennett, Daniel. 1995. *Darwin's Dangerous Idea: Evolution and the Meanings of Life.* London: Allen Lane/The Penguin Press.

Diamond, Jared. 1997. *Guns, Germs, and Steel: The Fates of Human Societies.* New York: W. W. Norton & Company.

Dines, Gail, Robert Jensen, and Ann Russo. 1998. *Pornography: The Production and Consumption of Inequality.* London: Routledge.

Donovan, Josephine. 2000. *Feminist Theory: The Intellectual Traditions*, 3rd ed. New York: Continuum.

Dupré, John. 2001. Evolution and Gender. *Women: A Cultural Review* 12(1):9–18.

Durkheim, Emile 1958 [1895]. *The Rules of Sociological Method.* Glencoe: Free Press.

Dusek, Val. 1998. Sociobiology Sanitized: The Evolutionary Psychology and Genic Selectionism Debates. *Science and Culture*, www.shef.ac.uk/~psysc/rmy/dusek.html.

Dworkin, Andrea. 1997. *Life and Death: Unapologetic Writings on the Continuing War against Women.* New York: The Free Press.

Eagly, Alice. 1995. The Science and Politics of Comparing Women and Men. *American Psychologist* 50(3):145–158.

Eagly, Alice, and Wendy Wood. 1999. The Origins of Sex Differences in Human Behavior: Evolved Dispositions versus Social Roles. *American Psychologist* 54(6):408–423.

Ehrenreich, Barbara, and Janet McIntosh. 1997. The New Creationism: Biology under Attack. *The Nation*, September 6.

Ellis, Bruce. 1992. The Evolution of Sexual Attraction: Evaluative Mechanisms in Women. In *The Adapted Mind: Evolutionary Psychology and the Generation of Culture*, eds. Jerome Barkow, Leda Cosmides, and John Tooby. New York: Oxford University Press.

Ellis, Bruce, and Timothy Ketelaar. 2000. On the Natural Selection of Alternative Models: Evaluation of Explanations in Evolutionary Psychology. *Psychological Inquiry* 11(1):56–68.

Ellis, Bruce, and Donald Symons. 1990. Sex Differences in Sexual Fantasy: An Evolutionary Psychological Approach. In *Human Nature: A Critical Reader*, ed. Laura Betzig. New York: Oxford University Press.

Ellis, Lee. 1989. *Theories of Rape: Inquiries into the Causes of Sexual Aggression.* New York: Hemisphere.

Etcoff, Nancy. 1999. *Survival of the Prettiest: The Science of Beauty.* London: Little, Brown & Company.

Evans, Dylan. 2001. *Emotion: The Science of Sentiment.* New York: Oxford University Press.

Fausto-Sterling, Anne. 1992. *Myths of Gender: Biological Theories about Women and Men*, 2nd ed. New York: Basic Books.

———. 1993. The Five Sexes: Why Male and Female Are Not Enough. *The Sciences* 33(2):20–24.

———. 1995. Attacking Feminism Is No Substitute for Good Scholarship. *Politics and the Life Sciences* 14(2):171–174.

———. 1997. Feminism and Behavioral Evolution: A Taxonomy. In *Feminism and Evolutionary Biology: Boundaries, Intersections, and Frontiers*, ed. Patricia Gowaty. New York: Chapman & Hall.

———. 2000a. Beyond Difference: Feminism and Evolutionary Biology. In *Alas, Poor Darwin: Arguments against Evolutionary Psychology*, eds. Hilary Rose and Steven Rose. London: Jonathan Cape.

———. 2000b. *Sexing the Body: Gender Politics and the Construction of Sexuality.* New York: Basic Books.

Fedigan, Linda. 1997. Is Primatology a Feminist Science? In *Women in Human Evolution*, ed. Lori Hager. London: Routledge.

Fisher, Helen. 1999. *The First Sex: The Natural Talents of Women and How They Will Change the World.* New York: Random House Trade.

Fishman, Loren. 1996. Feelings and Beliefs. In *The Flight from Science and Reason*, eds. Paul Gross, Norman Levitt, and Martin Lewis. New York: The New York Academy of Sciences.

Fletcher, Garth. 2000. Evaluating Scientific Theories. *Psychological Inquiry* 11(1):29–31.

Fodor, Jerry. 1983. *The Modularity of Mind.* Cambridge: MIT Press.

Fox Keller, Evelyn. 1982. Feminism and Science. In *Sex and Scientific Inquiry*, eds. Sandra Harding and Jean O'Barr. Chicago: University of Chicago Press.

Fraser, Nancy, and Linda Nicholson. 1988. Social Criticism without Philosophy: An Encounter between Feminism and Postmodernism. In *Feminism/Postmodernism*, ed. Linda Nicholson. London: Routledge.

Freeman, Derek. 1983. *Margaret Mead and Samoa: The Making and Unmaking of an Anthropological Myth.* Cambridge: Harvard University Press.

French, Marilyn. 1992. *The War against Women.* London: Penguin Books.

Friedan, Betty. 1963. *The Feminine Mystique.* New York: Norton.

Fuller, Margaret. 1845. *Woman in the Nineteenth Century.* New York: W. W. Norton & Company, 1971.

Gamble, Eliza Burt. 1894. *The Evolution of Woman: An Inquiry into the Dogma of Her Inferiority to Man.* New York: G. P. Putnam's Sons.

Gardiner, Judith. 2002. Feminism and Psychoanalysis. In *The Freud Encyclopaedia: Theory, Therapy, and Culture*, ed. Edward Erwin. New York: Routledge.
Geary, David. 1998. *Male, Female: The Evolution of Human Sex Differences*. Washington, D.C.: American Psychological Association.
———. 2000. Evolution and Proximate Expression of Human Paternal Investment. *Psychological Bulletin* 126(1):55–77.
Gijs, Luk. 2002. Etiologische theorieën over seksueel agressief gedrag: Een inleidend overzicht. *Tijdschrift voor Seksuologie* 26:9–25.
Gilligan, Carol. 1982. *In a Different Voice: Psychological Theory and Women's Development*. Cambridge: Harvard University Press.
Goodall, Jane. 1999. *Reason for Hope: A Spiritual Journey*. New York: Warner.
Gottschall, Jonathan, and Tiffany Gottschall. 2003. Are Per-incident Rape-pregnancy Rates Higher than Pre-incident Consensual Pregnancy Rates? *Human Nature* 14(1):1–20.
Gould, Stephen J. 1997. Evolution: The Pleasures of Pluralism. *New York Review of Books*, June 26.
Gould, Stephen J., and Steven Pinker. 1997. Evolutionary Psychology: An Exchange. *New York Review of Books*, October 9.
Gowaty, Patricia. 1992. Evolutionary Biology and Feminism. *Human Nature* 3(3):217–249.
———. 1995. False Criticisms of Sociobiology and Behavioral Ecology: Genetic Determinism, Untestability, and Inappropriate Comparisons. *Politics and the Life Sciences* 14(2):174–180.
———, ed. 1997a. *Feminism and Evolutionary Biology: Boundaries, Intersections, and Frontiers*. New York: Chapman & Hall.
———. 1997b. Sexual Dialectics, Sexual Selection, and Variation in Reproductive Behavior. In *Feminism and Evolutionary Biology: Boundaries, Intersections, and Frontiers*, ed. Patricia Gowaty. New York: Chapman & Hall.
———. 1997c. Introduction: Darwinian Feminists and Feminist Evolutionists. In *Feminism and Evolutionary Biology: Boundaries, Intersections, and Frontiers*, ed. Patricia Gowaty. New York: Chapman & Hall.
———. 2003. Power Asymmetries between the Sexes, Mate Preferences, and Components of Fitness. In *Evolution, Gender, and Rape*, ed. Cheryl Brown Travis. Cambridge: MIT Press.
Gray, Russell. 1997. "In the Belly of the Monster": Feminism, Developmental Systems, and Evolutionary Explanations. In *Feminism and Evolutionary Biology: Boundaries, Intersections, and Frontiers*, ed. Patricia Gowaty. New York: Chapman & Hall.
Gribbin, John, and Michael White. 1995. *Darwin: A Life in Science*. London: Simon & Schuster, 1997.
Gross, Barry. 1996. Flights of Fancy: Science, Reason, and Common Sense. In *The Flight from Science and Reason*, eds. Paul Gross, Norman Levitt, and Martin Lewis. New York: The New York Academy of Sciences.
Gross, Paul, and Norman Levitt. 1994. *Higher Superstitions: The Academic Left and Its Quarrels with Science*. Baltimore: Johns Hopkins University Press.

Grosz, Elizabeth. 1999. Darwin and Feminism: Preliminary Investigations for a Possible Alliance. *Australian Feminist Studies* 14(29):31–45.

Grünbaum, Adolf. 2002. Critique of Psychoanalysis. In *The Freud Encyclopaedia: Theory, Therapy, and Culture*, ed. Edward Erwin. New York: Routledge.

Haack, Susan. 1996a. Concern for Truth: What It Means, Why It Matters. In *The Flight from Science and Reason*, eds. Paul Gross, Norman Levitt, and Martin Lewis. New York: The New York Academy of Sciences.

———. 1996b. Towards a Sober Sociology of Science. In *The Flight from Science and Reason*, eds. Paul Gross, Norman Levitt, and Martin Lewis. New York: The New York Academy of Sciences.

Haig, Brian, and Russil Durrant. 2000. Theory Evaluation in Evolutionary Psychology. *Psychological Inquiry* 11(1):34–38.

Hall, Ruth. 2000. Reply to *A Natural History of Rape. The Independent*, February 23.

Hamilton, William, and Marlene Zuk. 1982. Heritable True Fitness and Bright Birds: A Role for Parasites? *Science* 218:384–387.

Haraway, Donna. 1991a. Animal Sociology and a Natural Economy of the Body Politic: A Political Physiology of Dominance. In *Simians, Cyborgs, and Women*. London: Free Association Books.

———. 1991b. In the Beginning Was the Word: The Genesis of Biological Theory. In *Simians, Cyborgs, and Women*. London: Free Association Books.

———. 1991c. The Contest for Primate Nature: Daughters of Man-the-Hunter in the Field, 1960–80. In *Simians, Cyborgs, and Women*. London: Free Association Books.

———. 1991d. A Cyborg Manifesto: Science, Technology, and Social-Feminism in the Late Twentieth Century. In *Simians, Cyborgs, and Women*. London: Free Association Books.

———. 1991e. *Simians, Cyborgs, and Women: The Reinvention of Nature*. London: Free Association Books.

———. 1991f. Situated Knowledges: The Science Question in Feminism and the Privilege of Partial Perspective. In *Simians, Cyborgs, and Women*. London: Free Association Books.

Harding, Cheryl. 1985. Sociobiological Hypotheses about Rape: A Critical Look at the Data behind the Hypotheses. In *Violence against Women: A Critique of the Sociobiology of Rape*, eds. Suzanne Sunday and Ethel Tobach. New York: Gordian Press.

Harding, Sandra. 1986. The Instabilities of the Analytical Categories of Feminist Theory. In *Sex and Scientific Inquiry*, eds. Sandra Harding and Jean O'Barr. Chicago: University of Chicago Press.

Harris, Judith Rich. 1998. *The Nurture Assumption: Why Children Turn Out the Way They Do*. New York: Free Press.

Hartmann, Heidi. 1979. The Unhappy Marriage of Marxism and Feminism: Towards a More Progressive Union. In *Women and Revolution*, ed. Lydia Sargent. London: South End Press.

Hermans, Cor. 2003. *De dwaaltocht van het sociaal-darwinisme. Vroege sociale interpretaties van Charles Darwins theorie van natuurlijke selectie 1859–1918*. Amsterdam: Nieuwezijds.

Hermsen, Joke. 1997. Vrouwenstudies Filosofie. *Filosofie* 7(5):4–7.

Hinde, Robert. 1987. Can Nonhuman Primates Help Us Understand Human Behavior? In *Primate Societies*, eds. Barbara Smuts et al. Chicago: University of Chicago Press.
Hoff Sommers, Christina. 1994. *Who Stole Feminism? How Women Have Betrayed Women*. New York: Touchstone, 1995.
———. 2000. *The War against Boys: How Misguided Feminism Is Harming Our Young Men*. New York: Simon & Schuster.
Holmes, Donna, and Christine Hitchcock. 1997. A Feeling for the Organism? An Empirical Look at Gender and Research Choices of Animal Behaviorists. In *Feminism and Evolutionary Biology: Boundaries, Intersections, and Frontiers*, ed. Patricia Gowaty. New York: Chapman & Hall.
hooks, bell. 1981. *Ain't I a Woman: Black Women and Feminism*. Boston: South End Press.
Hrdy, Sarah. 1997. Raising Darwin's Consciousness: Female Sexuality and the Prehominid Origins of Patriarchy. *Human Nature* 8(1):1–49.
———. 1999a. *The Woman That Never Evolved*, 2nd ed. Cambridge: Harvard University Press. (Originally published in 1981.)
———. 1999b. *Mother Nature: Natural Selection and the Female of the Species*. London: Chatto & Windus.
Hubbard, Ruth. 1988. Some Thoughts about the Masculinity of the Natural Sciences. In *Feminist Thought and the Structure of Knowledge*, ed. Mary Gergen. New York: New York University Press.
———. 1990. *The Politics of Women's Biology*. New Brunswick: Rutgers University Press.
Humphrey, Nicholas. 1976. The Social Function of the Intellect. In *Growing Points in Ethology*, eds. P. P. G. Bateson and R. A. Hinde. New York: Cambridge University Press.
———. 1992. *A History of the Mind*. London: Vintage.
———. 2000. How to Solve the Mind-Body Problem. *Journal of Consciousness Studies* 7(4):5–20.
Hyde, Janet. 1996. Where Are the Gender Differences? Where Are the Gender Similarities? In *Sex, Power, Conflict: Evolutionary and Feminist Perspectives*, eds. David Buss and Neil Malamuth. New York: Oxford University Press.
Israëls, Han. 1999. *De Weense kwakzalver. Honderd jaar Freud en de freudianen*. Amsterdam: Bert Bakker.
Jacob, François. 1997. *Of Flies, Mice, and Men*. Cambridge: Harvard University Press.
Jeffreys, Sheila. 1990. *Anticlimax: A Feminist Perspective on the Sexual Revolution*. Berkeley: University of California Press.
Jones, Owen. 1999. Sex, Culture, and the Biology of Rape: Toward Explanation and Prevention. *California Law Review* 87(4):827–942.
———. 2000. Law and the Biology of Rape: Reflections on Transitions. *Hastings Women's Law Journal* 11(2):151–178.
———. 2001. Realities of Rape: Of Science and Politics, Causes and Meanings. *Cornell Law Review* 86(6):1386–1422.
Kandel, Eric. 1999. Biology and the Future of Psychoanalysis: A New Intellectual Framework for Psychiatry Revisited. *American Journal of Psychiatry* 156(4):505–524.
Kenrick, Douglas, Melanie Trost, and Virgil Sheets. 1996. Power, Harassment, and Trophy Mates: The Feminist Advantages of an Evolutionary Perspective. In *Sex,*

Power, Conflict: Evolutionary and Feminist Perspectives, eds. David Buss and Neil Malamuth. New York: Oxford University Press.

Ketelaar, Timothy, and Bruce Ellis. 2000. Are Evolutionary Explanations Unfalsifiable? Evolutionary Psychology and the Lakatosian Philosophy of Science. *Psychological Inquiry* 11(1):1–21.

Kimmel, Michael. 2003. An Unnatural History of Rape. In *Evolution, Gender, and Rape*, ed. Cheryl Brown Travis. Cambridge: MIT Press.

Kimura, Doreen. 1999. *Sex and Cognition.* Cambridge: MIT Press, 2000.

———. 2002. Sex Hormones Influence Human Cognitive Pattern. *Neuroendocrinology Letters* 23:67–77.

Knapp, A. Bernard. 1998. Boys Will Be Boys: Masculinist Approaches to a Gendered Archaeology. In *Reader in Gender Archeology*, eds. Kelley Hays-Gilpin and David Whitley. London: Routledge.

Knight, Jonathan. 2002. Sexual Stereotypes. *Nature* 415, January 17, 254–256.

Koertge, Noretta. 1996a. Wrestling with the Social Constructor. In *The Flight from Science and Reason*, eds. Paul Gross, Norman Levitt, and Martin Lewis. New York: The New York Academy of Sciences.

———. 1996b. Feminist Epistemology: Stalking an Un-Dead Horse. In *The Flight from Science and Reason*, eds. Paul Gross, Norman Levitt, and Martin Lewis. New York: The New York Academy of Sciences.

Koyama, Nicola, Andrew McGain, and Russell Hill. 2004. Self-Reported Mate Preferences and "Feminist" Attitudes Regarding Marital Relations. *Evolution and Human Behavior* 25(5):327–335.

Kuhn, Thomas. 1996. *The Structure of Scientific Revolutions*, 3rd ed. Chicago: University of Chicago Press. (Originally published in 1962.)

Kurzban, Robert. 2002. Alas Evolutionary Psychology: Unfairly Accused, Unjustly Condemned. *The Human Nature Review* 2:99–109.

Laqueur, Thomas. 1990. *Making Sex: Body and Gender from the Greeks to Freud.* Cambridge: Harvard University Press.

Lawton, Marcy, William Garstka, and J. Craig Hanks. 1997. The Mask of Theory and the Face of Nature. In *Feminism and Evolutionary Biology: Boundaries, Intersections, and Frontiers*, ed. Patricia Gowaty. New York: Chapman & Hall.

Lenington, Sarah. 1985. Sociobiological Theory and the Violent Abuse of Women. In *Violence against Women: A Critique of the Sociobiology of Rape*, eds. Suzanne Sunday and Ethel Tobach. New York: Gordian Press.

Lewis, Martin. 1996. Radical Environmental Philosophy and the Assault on Reason. In *The Flight from Science and Reason*, eds. Paul Gross, Norman Levitt, and Martin Lewis. New York: The New York Academy of Sciences.

Liesen, Laurette. 1995a. Feminism and the Politics of Reproductive Strategies. *Politics and the Life Sciences* 14(2):145–162.

———. 1995b. Beyond the Dualism of Nature and Culture: Sex and the Frontiers of Human Nature Theory. *Politics and the Life Sciences* 14(2):194–197.

———. 1998. The Legacy of Woman the Gatherer: The Emergence of Evolutionary Feminism. *Evolutionary Anthropology* 7(3):105–113.

Lopreato, Joseph, and Timothy Crippen. 1999. *Crisis in Sociology: The Need for Darwin.* New Jersey: Transaction Publishers.

Lorenz, Konrad. 1965. *Evolution and Modification of Behavior.* Chicago: University of Chicago Press.

Low, Bobbi. 1989. Cross-Cultural Patterns in the Training of Children: An Evolutionary Perspective. In *Human Nature: A Critical Reader*, ed. Laura Betzig. New York: Oxford University Press.

———. 1992. Men, Women, Resources, and Politics in Pre-Industrial Societies. In *The Nature of the Sexes: The Sociobiology of Sex Differences and the "Battle of the Sexes,"* ed. Johan van der Dennen. Groningen: Origin Press.

———. 2000. *Why Sex Matters: A Darwinian Look at Human Behavior.* Princeton: Princeton University Press.

Malamuth, Neil. 1996. The Confluence Model of Sexual Aggression: Feminist and Evolutionary Perspectives. In *Sex, Power, Conflict: Evolutionary and Feminist Perspectives*, eds. David Buss and Neil Malamuth. New York: Oxford University Press.

Malthus, Thomas. 1976 [1798]. *An Essay on the Principle of Population.* New York/London: W.W. Norton & Company.

Margolin, Leslie, and Lynn White. 1987. The Continuing Role of Physical Attractiveness in Marriage. *Journal of Marriage and the Family* 49:21–27.

Maynard Smith, John. 1997. Commentary. In *Feminism and Evolutionary Biology: Boundaries, Intersections, and Frontiers*, ed. Patricia Gowaty. New York: Chapman & Hall.

Mead, Margaret. 1928. *Coming of Age in Samoa: A Psychological Study of Primitive Youth for Western Civilisation.* New York: Blue Ribbon Books.

———. 1935. *Sex and Temperament in Three Primitive Societies.* New York: William Morrow.

Mealey, Linda. 2000. *Sex Differences: Developmental and Evolutionary Strategies.* San Diego: Academic Press.

Merchant, Carolyn. 1980. *The Death of Nature: Women, Ecology, and the Scientific Revolution.* New York: HarperCollins Publishers.

Mesnick, Sarah. 1997. Sexual Alliances: Evidence and Evolutionary Implications. In *Feminism and Evolutionary Biology: Boundaries, Intersections, and Frontiers*, ed. Patricia Gowaty. New York: Chapman & Hall.

Michielsens, Magda, Dimitri Mortelmans, Sonja Spee, and Mic Billet, eds. 1999. *Bouw een vrouw. Sociale constructie van vrouwbeelden in de media.* Gent: Academia Press.

Miller, Geoffrey. 2000a. How to Keep Our Metatheories Adaptive: Beyond Cosmides, Tooby, and Lakatos. *Psychological Inquiry* 11(1):42–46.

———. 2000b. *The Mating Mind: How Sexual Choice Shaped the Evolution of Human Nature.* London: Vintage.

Millett, Kate. 1970. *Sexual Politics.* London: Virago.

Mitchell, Juliet. 1974. *Psychoanalysis and Feminism.* London: Allen Lane.

Mithen, Steven. 1996. *The Prehistory of the Mind: A Search for the Origins of Art, Religion, and Science.* London: Phoenix.

Moller Okin, Susan. 1999. Is Multiculturalism Bad for Women? In *Is Multiculturalism Bad for Women? Susan Moller Okin with Respondents*, eds. Joshua Cohen, Matthew Howard, and Martha Nussbaum. Princeton: Princeton University Press.

Morris, Desmond. 1967. *The Naked Ape.* New York: Dell.

Muehlenhard, Charlene, Sharon Danoff-Burg, and Irene Powch. 1996. Is Rape Sex or Violence? Conceptual Issues and Implications. In *Sex, Power, Conflict: Evolutionary and Feminist Perspectives*, eds. David Buss and Neil Malamuth. New York: Oxford University Press.

Mysterud, Iver. 1996. Communicating Ideas about Humans and Evolutionary Theory. *Trends in Ecological Evolution* 11:310.

Nanda, Meera. 1996. The Science Question in Postcolonial Feminism. In *The Flight from Science and Reason*, eds. Paul Gross, Norman Levitt, and Martin Lewis. New York: The New York Academy of Sciences.

Nelissen, Mark. 2000. *De bril van Darwin. Op zoek naar de wortels van ons gedrag*. Tielt: Lannoo.

Nelkin, Dorothy. 2000. Less Selfish than Sacred? Genes and the Religious Impulse in Evolutionary Psychology. In *Alas, Poor Darwin: Arguments against Evolutionary Psychology*. London: Jonathan Cape.

Nicholson, Linda. 1994. Interpreting Gender. *Signs* 11:79–105.

Nicolson, Paula. 2000. Barriers to Women's Success: Are They Natural or Man-Made? *Psychology, Evolution, and Gender* 2(1):91–96.

Noske, Barbara. 1989. *Humans and Other Animals: Beyond the Boundaries of Anthropology*. London: Pluto Press.

Oliver, Mary Beth, and Janet Hyde. 1993. Gender Differences in Sexuality: A Meta-Analysis. *Psychological Bulletin* 114:29–51.

Palmer, Craig, and Randy Thornhill. 2003a. A Posse of Good Citizens Brings Outlaw Evolutionists to Justice. A Response to *Evolution, Gender, and Rape*. *Evolutionary Psychology* 1:10–27.

———. 2003b. Straw Men and Fairy Tales: Evaluating Reactions to *A Natural History of Rape*. *Journal of Sex Research* 40(3):249–255.

Patai, Daphne. 1998. *Heterophobia: Sexual Harassment and the Future of Feminism*. Lanham, Md.: Rowman & Littlefield.

Paul, Diane. 2002. Darwin, Social Darwinism and Eugenics. In *The Cambridge Companion to Darwin*, eds. Jonathan Hodge and Gregory Radick. Cambridge: Cambridge University Press.

Peplau, Letitia Anne. 2003. Human Sexuality: How Do Men and Women Differ? *Current Directions in Psychological Research* 12(2):37.

Peplau, Letitia Anne, and Linda Garnets. 2000. A New Paradigm for Understanding Women's Sexuality and Sexual Orientation. *Journal of Social Issues* 56(2):330–350.

Phillips, Helen. 2001. Gender: Why Two Sexes Are Not Enough. *New Scientist*, 2290:27–41.

Pinker, Steven. 1997. *How the Mind Works*. London: Penguin Books.

———. 2002. *The Blank Slate: The Modern Denial of Human Nature*. London: Allen Lane/The Penguin Press.

Pinker, Steven, and Paul Bloom. 1992. Natural Language and Natural Selection. In *The Adapted Mind: Evolutionary Psychology and the Generation of Culture*, eds. Jerome Barkow, Leda Cosmides, and John Tooby. New York: Oxford University Press.

Plotkin, Henry. 1997. *Evolution in Mind: An Introduction to Evolutionary Psychology*. London: Penguin Books.

Popper, Karl. 1959. *The Logic of Scientific Discovery.* New York: Basic Books.

Pratto, Felicia. 1996. Sexual Politics: The Gender Gap in the Bedroom, the Cupboard, and the Cabinet. In *Sex, Power, Conflict: Evolutionary and Feminist Perspectives*, eds. David Buss and Neil Malamuth. New York: Oxford University Press.

Primoratz, Igor. 1999. *Ethics and Sex.* London: Routledge.

Profet, Margie. 1992. Pregnancy Sickness as Adaptation: A Deterrent to Maternal Ingestion of Teratogens. In *The Adapted Mind: Evolutionary Psychology and the Generation of Culture*, eds. Jerome Barkow, Leda Cosmides, and John Tooby. New York: Oxford University Press.

Radcliffe Richards, Janet. 1995. Why Feminist Epistemology Isn't (and the Implications for Feminist Jurisprudence). *Legal Theory* 1:365–400.

———. 1997. Equality of Opportunity. *Ratio* 10(3):253–279.

———. 1998. Feminism and Equality. *Journal of Contemporary Legal Issues* 9:225–247.

———. 2000. *Human Nature after Darwin: A Philosophical Introduction.* London: Routledge.

Rees, Amanda. 2000. Higamous, Hogamous, Woman Monogamous. *Feminist Theory* 1(3):365–370.

Rich, Madeline. 2000. Violent Attacks. *Chicago Tribune*, February 1.

Ridley, Matt. 1999. *Genome: The Autobiography of a Species in 23 Chapters.* New York: HarperCollins Publishers.

———. 2003. *Nature via Nurture: Genes, Experience, and What Makes Us Human.* London: Fourth Estate.

Roele, Marcel. 2000. *De mietjesmaatschappij. Over politiek incorrecte feiten.* Amsterdam: Contact.

Roiphe, Katie. 1993. *The Morning after: Sex, Fear, and Feminism on Campus.* Boston: Little, Brown.

Roscoe, Will. 1991. *The Zuni Man-Woman.* Albuquerque: University of New Mexico Press.

Rose, Hilary. 1983. Hand, Brain, and Heart: A Feminist Epistemology for the Natural Sciences. In *Sex and Scientific Inquiry*, eds. Sandra Harding and Jean O'Barr. Chicago: University of Chicago Press.

———. 1997. Beyond Biology. *Red Pepper*, September.

———. 2000. Colonising the Social Sciences? In *Alas, Poor Darwin: Arguments against Evolutionary Psychology*, eds. Hilary Rose and Steven Rose. London: Jonathan Cape.

Rose, Hilary, and Steven Rose, eds. 2000a. *Alas, Poor Darwin: Arguments against Evolutionary Psychology.* London: Jonathan Cape.

———. 2000b. Introduction. In *Alas, Poor Darwin: Arguments against Evolutionary Psychology*, eds. Hilary Rose and Steven Rose. London: Jonathan Cape.

Rose, Steven. 2000. The New Just So Stories: Sexual Selection and the Fallacies of Evolutionary Psychology. *Times Literary Supplement*, July 14.

Rosser, Sue. 1992. *Biology and Feminism: A Dynamic Interaction.* New York: Twayne Publishers.

———. 1997. Possible Implications of Feminist Theories for the Study of Evolution. In *Feminism and Evolutionary Biology: Boundaries, Intersections, and Frontiers*, ed. Patricia Gowaty. New York: Chapman & Hall.

———. 2003. Coming Full Circle: Refuting Biological Determinism. In *Evolution, Gender, and Rape*, ed. Cheryl Brown Travis. Cambridge and London: MIT Press.

Ruse, Michael. 1998. Is Darwinism Sexist? (And If It Is, So What?). In *A House Built on Sand: Exposing Postmodernist Myths about Science*, ed. Noretta Koertge. New York: Oxford University Press.

Ruskai, Mary Beth. 1996. Are "Feminist Perspectives" in Mathematics and Science Feminist? In *The Flight from Science and Reason*, eds. Paul Gross, Norman Levitt, and Martin Lewis. New York: The New York Academy of Sciences.

Russell, K. P. 1977. *Eastman's Expectant Motherhood.* 6th ed. New York: Little.

Sanday, Peggy Reeves. 1986. Rape and the Silencing of the Feminine. In *Rape*, eds. Sylvana Tomaselli and Roy Porter. Oxford: Blackwell.

———. 2003. Rape-Free versus Rape-Prone: How Culture Makes a Difference. In *Evolution, Gender, and Rape*, ed. Cheryl Brown Travis. Cambridge: MIT Press.

Sanger, Alexander. 2004. *Beyond Choice: Reproductive Freedom in the 21st Century.* New York: Public Affairs.

Sapiro, Virginia. 1985. Biology and Women's Policy: A View from the Social Sciences. In *Women, Biology, and Public Policy*, ed. Virginia Sapiro. Beverly Hills, Calif.: Sage Publications.

Sax, Leonard. 2002. How Common Is Intersex? A Response to Anne Fausto-Sterling. *Journal of Sex Research* 39(3):174–178.

Schiebinger, Londa L. 1999. *Has Feminism Changed Science?* Cambridge: Harvard University Press.

Schmitt, David, et al. 2003a. Universal Sex Differences in the Desire for Sexual Variety: Tests from 52 nations, 6 Continents, and 13 Islands. *Journal of Personality and Social Psychology* 85(1):85–104.

———. 2003b. Are Men Universally More Dismissing than Women? Gender Differences in Romantic Attachment across 62 Cultural Regions. *Personal Relationships* 10(3):307–331.

———. 2004. Patterns and Universals of Mate Poaching across 53 Nations: The Effects of Sex, Culture, and Personality on Romantically Attracting Another Person's Partner. *Journal of Personality and Social Psychology* 86(4):560–584.

Schwendinger, Julia, and Herman Schwendinger. 1985. Homo Economicus as the Rapist in Sociobiology. In *Violence against Women: A Critique of the Sociobiology of Rape*, eds. Suzanne Sunday and Ethel Tobach. New York: Gordian Press.

Segal, Lynne. 1999. *Why Feminism?* New York: Columbia University Press.

Segal, Lynne, and Simon Baron-Cohen. 2003. Sex on the Mind: An Email Exchange. *The Guardian*, May 3.

Segerstråle, Ullica. 1992. Sociobiology and Feminism: Enemies or Allies? In *The Nature of the Sexes: The Sociobiology of Sex Differences and the "Battle of the Sexes,"* ed. Johan van der Dennen. Groningen: Origin Press.

———. 2000. *Defenders of the Truth: The Battle for Science in the Sociobiology Debate and Beyond.* New York: Oxford University Press.

Shakespeare, Tom, and Mark Erickson. 2000. Different Strokes: Beyond Biological Determinism and Social Constructionism. In *Alas, Poor Darwin: Arguments against Evolutionary Psychology.* London: Jonathan Cape.

Sheehan, Elizabeth. 1997. Victorian Clitoridectomy: Isaac Baker Brown and His Harmless Little Procedure. In *The Gender/Sexuality Reader*, eds. Roger Lancaster and Micaela di Leonardo. New York: Routledge.

Shermer, Michael. 2001. *The Borderlands of Science: Where Sense Meets Nonsense.* New York: Oxford University Press.

Shields, Stephanie, and Pamela Steinke. 2003. Does Self-Report Make Sense as an Investigative Method in Evolutionary Psychology? In *Evolution, Gender, and Rape*, ed. Cheryl Brown Travis. Cambridge: MIT Press.

Shields, William, and Lea Shields. 1983. Forcible Rape: An Evolutionary Perspective. *Ethology and Sociobiology* 4:115–136.

Silverman, Irwin, and Marion Eals. 1992. Sex Differences in Spatial Abilities: Evolutionary Theory and Data. In *The Adapted Mind: Evolutionary Psychology and the Generation of Culture*, eds. Jerome Barkow, Leda Cosmides, and John Tooby. New York: Oxford University Press.

Singer, Peter. 1999. *A Darwinian Left: Politics, Evolution and Cooperation.* London: Weidenfeld & Nicolson.

Singh, Devendra. 1993. Adaptive Significance of Female Physical Attractiveness: Role of Waist-to-Hip Ratio. *Journal of Personality and Social Psychology* 65(2):293–307.

———. 2002. Female Mate Value at a Glance: Relationship of Waist-to-Hip Ratio to Health, Fecundity, and Attractiveness. *Neuroendocrinology Letters* 23:81–91.

Smith, Eric, Monique Borgerhoff Mulder, and Kim Hill. 2001. Controversies in the Evolutionary Social Sciences: A Guide for the Perplexed. *Trends in Ecology and Evolution* 16(3):128–135.

Smuts, Barbara. 1995. The Evolutionary Origins of Patriarchy. *Human Nature* 6(1):1–32.

———. 1996. Male Aggression against Women: An Evolutionary Perspective. In *Sex, Power, Conflict: Evolutionary and Feminist* Perspectives, eds. David Buss and Neil Malamuth. New York: Oxford University Press.

Snowdon, Charles. 1997. The "Nature" of Sex Differences: Myths of Male and Female. In *Feminism and Evolutionary Biology: Boundaries, Intersections, and Frontiers*, ed. Patricia Gowaty. New York: Chapman & Hall.

Sork, Victoria. 1997. Quantitative Genetics, Feminism, and Evolutionary Theories of Gender Differences. In *Feminism and Evolutionary Biology: Boundaries, Intersections, and Frontiers*, ed. Patricia Gowaty. New York: Chapman & Hall.

Spaink, Karin. 1994. Cyborgs zijn heel gewone mensen (ze denken hooguit meer na). In Donna Haraway, *Een Cyborg Manifest.* Amsterdam: De Balie.

Speelman, Tom, Johan Braeckman, and Griet Vandermassen. 2001. De modulen van de geest. Misverstanden over evolutiepsychologie. *Skepter* 14(4):29–32.

Stanton, Elizabeth C., Susan B. Anthony, and Matilda J. Gage, eds. 1881/1882. Selections from the *History of Woman Suffrage.* In *The Feminist Papers*, ed. Alice Rossi. New York: Bantam Books.

Steiner, George 1971. *In Bluebeard's Castle.* New Haven, Conn.: Yale University Press.

Stengers, Isabelle. 1997. *Macht en wetenschappen.* Brussel: VUBPress.

Storr, Anthony. 1989. *Freud.* New York: Oxford University Press.

Sunday, Suzanne, and Ethel Tobach, eds. 1985. *Violence against Women: A Critique of the Sociobiology of Rape.* New York: Gordian Press.

Symons, Donald. 1992. On the Use and Misuse of Darwinism in the Study of Human Behavior. In *The Adapted Mind: Evolutionary Psychology and the Generation of Culture*, eds. Jerome Barkow, Leda Cosmides, and John Tooby. New York: Oxford University Press.

———. 1995. Beauty Is in the Adaptations of the Beholder: The Evolutionary Psychology of Human Female Sexual Attractiveness. In *Sexual Nature/Sexual Culture*, eds. Paul Abramson and Steven Pinkerton. Chicago: University of Chicago Press.

Symons, Donald, Catherine Salmon, and Bruce Ellis. 1997. Unobtrusive Measures of Human Sexuality. In *Human Nature: A Critical Reader*, ed. Laura Betzig. New York: Oxford University Press.

Talarico, Susette M. 1985. An Analysis of Biosocial Theories of Crime. In *Women, Biology and Public Policy*, ed. Virginia Sapiro. Beverly Hills, Calif.: Sage Publications.

Tang-Martinez, Zuleyma. 1997. The Curious Courtship of Sociobiology and Feminism: A Case of Irreconcilable Differences. In *Feminism and Evolutionary Biology: Boundaries, Intersections, and Frontiers*, ed. Patricia Gowaty. New York: Chapman & Hall.

ten Dam, Geert, and Monique Volman. 1995. Continuïteit en verandering. In *Vrouwenstudies in de jaren negentig. Een kennismaking vanuit verschillende disciplines*, eds. Margo Brouns et al. Bussum: Coutinho.

Thornhill, Nancy, and Randy Thornhill. 1990. An Evolutionary Analysis of Psychological Pain Following Rape: I. The Effects of Victim's Age and Marital Status. *Ethology and Sociobiology* 11:155–176.

Thornhill, Randy, and Craig Palmer. 2000. *A Natural History of Rape: Biological Bases of Sexual Coercion*. Cambridge: MIT Press.

———. 2001. Rape and Evolution: A Reply to Our Critics. mitpress.mit.edu/books/THOUH/thornhill-preface.pdf.

Thornhill, Randy, and Nancy Thornhill. 1983. Human Rape: An Evolutionary Analysis. *Ethology and Sociobiology* 4:137–173.

Thornhill, Randy, Nancy Thornhill, and Gerard Dizinno. 1986. The Biology of Rape. In *Rape*, eds. Sylvana Tomaselli and Roy Porter. Oxford: Blackwell.

Tobach, Ethel, and Rachel Reed. 2003. Understanding Rape. In *Evolution, Gender, and Rape*, ed. Cheryl Brown Travis. Cambridge: MIT Press.

Tobach, Ethel, and Suzanne Sunday. 1985. Epilogue. In *Violence against Women: A Critique of the Sociobiology of Rape*, eds. Suzanne Sunday and Ethel Tobach. New York: Gordian Press.

Tomaselli, Sylvana, and Roy Porter, eds. 1986. *Rape*. Oxford: Blackwell.

Tooby, John, and Leda Cosmides. 1992. The Psychological Foundations of Culture. In *The Adapted Mind: Evolutionary Psychology and the Generation of Culture*, eds. Jerome Barkow, Leda Cosmides, and John Tooby. New York: Oxford University Press.

Torrey, E. Fuller. 1992. *Freudian Fraud: The Malignant Effect of Freud's Theory on American Thought and Culture*. New York: HarperPerennial.

Travis, Cheryl Brown, ed. 2003a. *Evolution, Gender, and Rape*. Cambridge: MIT Press.

———. 2003b. Talking Evolution and Selling Difference. In *Evolution, Gender, and Rape*, ed. Cheryl Brown Travis. Cambridge: MIT Press.

———. 2003c. Theory and Data on Rape and Evolution. In *Evolution, Gender, and Rape*, ed. Cheryl Brown Travis. Cambridge: MIT Press.

Trivers, Robert. 1972. Parental Investment and Sexual Selection. In *Sexual Selection and the Descent of Man, 1871–1971*, ed. Bernard Campbell. Chicago: Aldine.

Tuana, Nancy. 1988. The Weaker Seed: The Sexist Bias of Reproductive Theory. In *Feminism and Science*, ed. Nancy Tuana. Bloomington: Indiana University Press.

van der Dennen, Johan. 2002. (Evolutionary) Theories of Warfare in Preindustrial (Foraging) Societies. *Neuroendocrinology Letters* 23:55–65.

van Muijlwijk, Margreet. 1998. *De toekomst van Teiresias. Vrouwelijke gestalten van het gemis.* Brussel: VUBPress.

Vandermassen, Griet. 2003. Revolutionaire psychologie. *Mores* 48(237):540–551.

———. 2004. Sexual Selection: A Tale of Male Bias and Feminist Denial. *European Journal of Women's Studies* 11(1):9–26.

Volscho, Thomas, and Robert Pietrzak. 2002. Do Men Have More Sex Partners than Women? Poster presented at Human Behavior and Evolution Society Conference, New Brunswick, N.J., June 19–23.

Waage, Jonathan. 1997. Parental Investment—Minding the Kids or Keeping Control? In *Feminism and Evolutionary Biology: Boundaries, Intersections, and Frontiers*, ed. Patricia Gowaty. New York: Chapman & Hall.

Waage, Jonathan, and Patricia Gowaty. 1997. Myths of Genetic Determinism. In *Feminism and Evolutionary Biology: Boundaries, Intersections, and Frontiers*, ed. Patricia Gowaty. New York: Chapman & Hall.

Watson, John. 1913. Psychology as the Behaviorist Views It. *Psychological Review* 20:158–177.

Watson-Brown, Linda. 2000. Foolish Findings of the Doctor Who Believes Rapists Are Born Evil. *The Scotsman*, July 10.

Wennerås, Christine, and Agnes Wold. 1997. Nepotism and Sexism in Peer-Review. *Nature* 387:341–343.

Wertheim, Margaret. 2000. The Boy Can't Help It. *LA Weekly*, March 24.

Williams, George. 1966. *Adaptation and Natural Selection.* Princeton, N.J.: Princeton University Press.

Wilson, Edward. 1975. *Sociobiology: The New Synthesis.* Cambridge: The Belknapp Press of Harvard University Press.

———. 1978. *On Human Nature.* Cambridge: Harvard University Press.

Wilson, Elizabeth. 1999. Somatic Compliance—Feminism, Biology and Science. *Australian Feminist Studies* 14(29):7–18.

———. 2002. Biologically Inspired Feminism: Response to Helen Keane and Martha Rosengarten, "On the Biology of Sexed Subjects." *Australian Feminist Studies* 17(39):283–285.

Wilson, Margo, and Martin Daly. 1992. The Man Who Mistook His Wife for a Chattel. In *The Adapted Mind: Evolutionary Psychology and the Generation of Culture*, eds. Jerome Barkow, Leda Cosmides, and John Tooby. New York: Oxford University Press.

Wilson, Margo, and Sarah Mesnick. 1997. An Empirical Test of the Bodyguard Hypothesis. In *Feminism and Evolutionary Biology: Boundaries, Intersections, and Frontiers*, ed. Patricia Gowaty. New York: Chapman & Hall.

Wilson, Margo, Martin Daly, and Joanna Scheib. 1997. Femicide: An Evolutionary Psychological Perspective. In *Feminism and Evolutionary Biology: Boundaries, Intersections, and Frontiers*, ed. Patricia Gowaty. New York: Chapman & Hall.

Wollstonecraft, Mary. 1792. *A Vindication of the Rights of Woman.* New York: Random House.

Wood, Wendy, and Alice Eagly. 2000. A Call to Recognize the Breadth of Evolutionary Perspectives: Sociocultural Theories and Evolutionary Psychology. *Psychological Inquiry* 11(1):52–55.

Woolf, Virginia. 2000. *A Room of One's Own/Three Guineas.* London: Penguin Books. (Originally published in 1929 and 1938.)

Wright, Robert. 1990. The Intelligence Test. *New Republic*, January 29.

———. 1994a. Feminists, Meet Mr. Darwin. *New Republic*, November 28.

———. 1994b. *The Moral Animal: The New Science of Evolutionary Psychology.* London: Little, Brown & Company.

———. 1995. The Evolution of Despair. *Time*, August 28.

———. 2000. *Nonzero: The Logic of Human Destiny.* London: Little, Brown & Company.

Wylie, Alison. 1997. Good Science, Bad Science, or Science as Usual? Feminist Critiques of Science. In *Women in Human Evolution*, ed. Lori Hager. London: Routledge.

Young, Cathy. 1999. *Ceasefire! Why Women and Men Must Join Forces to Achieve True Equality.* New York: The Free Press.

———. 2003. Look Who's Cheating: Men, Women, and Evolution. *ReasonOnline*, www.reason.com/cy/cy081203.shtml.

Zihlman, Adrienne. 1997. The Paleolithic Glass Ceiling: Women in Human Evolution. In *Women in Human Evolution*, ed. Lori Hager. London: Routledge.

Zuk, Marlene. 1997. Darwinian Medicine Dawning in a Feminist Light. In *Feminism and Evolutionary Biology: Boundaries, Intersections, and Frontiers*, ed. Patricia Gowaty. New York: Chapman & Hall.

———. 2002. *Sexual Selections: What We Can and Can't Learn about Sex from Animals.* Berkeley/Los Angeles: University of California Press

Index

About the Author

Griet Vandermassen has an M.A. in philosophy and in philology. She is currently a Ph.D. student at the Centre for Gender Studies, University of Ghent, in Belgium. Before starting her academic career she worked as a freelance journalist. She has written or coauthored various articles in the fields of women's studies and evolutionary psychology. Her research interests include feminism, evolutionary theory, history of science, pseudo-science, and skepticism.